WHAT MAKES NATURE TICK?

ROGER G. NEWTON

WHAT MAKES
NATURE TICK?

HARVARD UNIVERSITY PRESS

CAMBRIDGE, MASSACHUSETTS

LONDON, ENGLAND 1993

Library of Congress Cataloging-in-Publication Data

Newton, Roger G.
  What makes nature tick? / Roger G. Newton.
    p. cm.
  Includes bibliographical references.
  ISBN 0-674-95085-2.
  1. Physics. I. Title.
  QC21.2.N53 1993
  530—dc20

93-9507
  CIP

To Lily, Isabella, and Eden
in the hope that they will grow up to love science

PREFACE

This book had its origin in a Phi Beta Kappa address and in a number of talks given over the last fifteen years to a group of colleagues from various disciplines who have been meeting at regular intervals to discuss our disparate intellectual interests. It is intended to be comprehensible to readers who are scientifically uneducated and who know very little mathematics. Such mathematics as is needed will be introduced at the appropriate places, not in detail sufficient to enable readers to solve problems, but to allow them to appreciate the role it plays in physics.

I am indebted to a number of people for inducing me to write this book and for helping me in that process. First and foremost, I am deeply grateful to my wife, Ruth, whose invaluable editorial assistance gives the work whatever pleasing style it has; any remaining clumsiness is entirely mine. I am also indebted to Professor Shehira Davezac of the Henry Radford Hope School of Fine Arts at Indiana University for assistance with the identification of works of art used for illustrations, to Danae Thimme of the Indiana University Art Museum for help with obtaining permission to copy works of art, and to many nonscientific colleagues who were my guinea-pig listeners when I tried to explain matters that I found fascinating.

Bloomington, May 1993   RGN

CONTENTS

WHAT MAKES NATURE TICK?

INTRODUCTION

Science is often distinguished from other human activities by the precision with which its practitioners formulate their statements. In 1948, when Julian Schwinger, using his newly constructed quantum electrodynamics to calculate the value of the "magnetic moment" of the electron, announced the result to be 1.001162 (in appropriate units), and compared it with the experimental value, which was then known to lie between 1.00115 and 1.00121, everyone had good reason to be impressed by an agreement between theory and experiment to a few parts in 100,000. (Over the last forty years the agreement has been refined to several parts in a billion.) For this work Schwinger, along with Richard Feynman and Sin-Itiro Tomonaga, received the Nobel Prize.

And yet it would be quite erroneous to think that statements made with great precision are necessarily scientific, while others made without any such claim are less so. When testing a theory against experimental measurements it is, of course, important to be able to carry precision as far as needed, which sometimes is very far indeed, using the intellectual and technical tools available to us in many areas of science. At other times, however, scientists are well satisfied with ball-park estimates, based on back-of-the-envelope calculations. Before an experimenter prepares or constructs an elaborate apparatus (or applies for the needed funds) to carry out a planned measurement, she has to have an idea of what to expect. Will there be a few counts per year, or several hundred thousand counts per second? Can the pointer on a dial be expected to move by a few thousandths of an inch or by three inches? The design of the instrument to be used will depend critically on such expectations, and rough estimates of predictions are as important for science as high-precision announcements; they are, in principle, no less scientific. "In thinking about the history of science in the period around sixty years ago, I have come to the conclusion," recently wrote Linus Pauling, one of the important contributors to the advance of physics and chemistry at that time, "that much of the progress was the result of carrying out *approximate* quantum mechanical calculations." He compares these favorably with other, much more precise computations that, in his opinion, provided very little physical insight. There are many circumstances when agreement to within 20% between theory and experiment is considered satisfactory, at least temporarily. On the other hand, a claim that brand X is "99.9% pure" gains no scientific stature by its pretense to precision, even though it tries. What is important for science is to appreciate the degree of precision that is appropriate and justified in a given circumstance.

It is, in fact, futile to try to characterize the nature of science within narrow limits, just as there is no rigidly identifiable scientific method; as the American physicist and quintessential Yankee Percy Bridgman put it, the scientific method is simply "to use your noodle and no holds barred." There are many kinds of science; some are mature, well-developed, and theoretically structured, others are young and still searching for the most appropriate tools. As for scientists, some of us are tinkerers, others classifiers; some are problem solvers, others conceptualizers; some ingenious and elegant, others thorough and meticulous. In addition, of course, some make important discoveries, moving the frontier of knowledge by perceptible amounts, while others fill in minor gaps to help complete a grand picture. But the one attribute those who make significant theoretical or experimental contributions must have is imagination. That is what makes science a *human* activity in the deepest sense.

The subject of this book is physics, the most mature of the sciences and the basic fund we have for understanding the workings of the universe. "Physics is an experimental science," its practitioners often proudly proclaim—and they are right. The ultimate basis of what we know of the universe and its functioning is based on observation of nature and the results of experiments. And yet this is only part of the truth. Observations alone are not sufficient to make up the body of knowledge we call science; they have to be filtered and digested by human intelligence. Indeed, the results of observations are meaningless without guidance and interpretation by theoretical concepts. Even the use of the most powerful computers for the classification of large bodies of collected data cannot substitute for the building of conceptual structures. The task of a creative experimenter is to ask nature the right questions, and that takes imagination as well as theoretical understanding of the implications of her previous answers. Theories are necessary not only to give meaning to "facts," but also to direct the coherent search for further knowledge.

The data that form the foundation of our knowledge in physics and chemistry are, for the most part (with the exception of astrophysics), obtained by active experimentation rather than passive observation. This stands in contrast to other sciences, such as the biological sciences and astronomy, in which experimentation sometimes is impossible, very difficult, or even unethical, and this difference has important consequences. As the long drawn-out controversies about the connection between smoking and lung cancer attest, causal links can be established much more persuasively by experimentation, in which some variables can be held fixed and others changed at will, than by straight observation. The reader should therefore keep in mind that when data are mentioned in this book, they are usually understood to be gathered by experiments.

Since imaginatively constructed theories form the essential underpinnings of a highly developed science, the question occasionally arises whether these constructs, and the facts observed under their guidance, may be influenced by extra-scientific considerations. A word that has therefore attracted a certain amount of controversy in discussions of science, both recently and in the past, is *objectivity*. Denying the stance of objectivity that scientists often adopt in discussions with nonscientists, some people claim that an objective attitude toward matters of fact is impossible, especially when these facts become of vital concern either to the researchers themselves or to society as a whole. There is, of course, some truth to this claim; scientists, after all, are human, and instances have come to light in which they were found to have been swayed in their judgment by their own interests, by ambition, or by their social contexts—either in priority disputes within science itself or in controversies concerning applica-

3

tions. But these human failures do not invalidate the ideal of objectivity—freedom from the influence of self-interest and socially or philosophically conditioned bias—an ideal that has always served science extremely well even when violated in practice.

The ideas of those who deny the validity of the goal of objectivity and even its desirability, usually on political grounds (whether in the name of Arian as opposed to Jewish mentality, materialist philosophy as opposed to bourgeois values, feminist against male patterns of thought, or Eastern intuition versus Western rationality), are fundamentally detrimental to science itself. While the thinking of scientists is undoubtedly, to some extent, subject to extra-scientific social, moral, and philosophical influences, these influences should not be exaggerated, and external attempts at steering or molding scientists' metaphysical preconceptions have always been futile at best and destructive at worst. The openness and accessibility of science make it to a remarkable degree self-correcting in the long run, and any existing bias will eventually be found out, so long as objectivity remains the guiding ideal.

This book is concerned primarily with the theoretical concepts that have been used in our attempts to understand the world around us over the last 400 years. We will encounter specific cases in which the originators of these theories were attacked on political grounds, or their ways of thinking were "explained" on the basis of their social surroundings. Such attacks and explanations are misplaced. The expressed thoughts of individuals may be suppressed or psychologically analyzed, but their scientific acceptance, in the long term, remains independent of both psychological origin and political success.

There are, nevertheless, valuable motivations at work in science that are extra-scientific and, to some extent, nonrational: these have to do with aesthetics. I hope the reader of this book will come away with an appreciation of the role beauty plays in the construction of scientific theories and the adoption of scientific concepts. Needless to say, beauty lies, to a large extent, in the eye of the beholder, and in some contexts aesthetic appreciation requires an educated taste. Nevertheless, there is a surprising amount of agreement among scientists in the same field about which theories are beautiful and which ugly.

Because the kind of education needed not only for comprehension but also for this refinement of taste is almost invariably mathematical, nonscientists often regard physics as inaccessible. I will attempt in this book to remove some of the barriers to an appreciation of the ideas and beauty of physics by introducing the needed mathematical notions in the text and developing some of them in accompanying boxes.

The first chapter expands the argument that, contrary to the manner in which science is frequently portrayed, it is not simply a collection of facts: though such facts are necessary, they do not constitute the most important part of science. What is more, very few physicists are motivated by the desire to be "relevant" or useful to society. Scientists require as much imagination and intuition for their work as artists, writers, and musicians. As the painter's medium is oil and brush, and the poets's is the music of his language, so the essential tool for a physicist's imagination is mathematics. While both in science and in mathematics beauty is a powerful motivating force, and our concepts are shaped at least in part by aesthetics, the fruits of a scientist's imagination must finally be rooted in quantitative agreement with experiments. Some of the general arguments of this chapter are applicable to all sciences, others are specific to the highly theoretically and mathematically structured science of physics.

Central to the second chapter is Laplace's famous claim: given enough information and intelligence, he would be able to predict the future course of the universe forever. To substantiate the basis of this statement, we begin by discussing the Newtonian equations of motion in classical mechanics, and in particular their form given by William Hamilton as first-order differential equations. Since the mechanics of point-like particles and rigid bodies is the most basic part of physics and the earliest to be fully developed, this is an appropriate place to start. We introduce the notion of phase space, in which the motion of a physical system is completely described as a unique curve through every point. Once one such point is given by the present state of a system, the trajectory is forever determined, just as Laplace proclaimed. However, almost all physical systems are "sensitive to initial conditions," which means that two systems whose initial states differ ever so slightly will eventually diverge in their behavior by such large amounts that they appear to have no resemblance to one another. Since small errors in the inital configuration of any physical system are practically unavoidable and, in addition, since any computation by digital computers necessarily introduces round-off errors, from a practical point of view, even with the aid of powerful computers, the behavior of most physical systems is not predictable in the long run—it is chaotic. The more particles the system contains, the sooner the "long run" takes over. As a result, Laplace's dream fades away, and only its *principle* remains. This chapter contains brief introductions to the concepts of vectors, of the differential calculus, and of ordinary differential equations. These introductions are formulated very concretely and are not intended to enable readers to solve problems, but only to familiarize them with the essential ideas.

The chaos introduced in the second chapter is enlarged upon in Chapter 3 for systems that consist of very many particles, such as gases and fluids, where chaos is most prevalent and is reached most quickly. We discuss the nature of heat and the ideas of statistical mechanics. This leads to the concept of entropy and the mysterious notion of irreversibility. A videotape of an egg falling and breaking, if run backwards, will immediately be recognized by any viewer as an impossible course of events. How does it come about that, even though each molecule follows the Newtonian equations of motion, which have no preference for one direction of time over the other—a viewer of a videotape of their motion cannot tell if it is running forward or backward—the behavior of a system containing a very large number of molecules acquires an "arrow of time"?

In Chapters 2 and 3, the forces acting among constituents of a physical system remained unexamined; in Chapter 4 we consider the nature of these forces. We start from Newton's action at a distance and his controversial long-range gravitational attraction, move to Faraday's invention of the field concept, and on to Maxwell, with his partial differential equations as the ultimate avoidance of action at a distance. (A brief introduction to the idea of a partial differential equation is given.) The carrier of the field is the *ether;* we have our first encounter with the experiment of Michelson and Morley, which was intended to find the speed of our motion in it. Einstein's general theory of relativity leads to the geometrization of the field. We also have a first view of the quantum theory, with photons and the quantum field. The concept of the field, central to modern theory, has been transformed: Feynman diagrams describe the action of "virtual particles," and the "polarization of the vacuum" makes even empty space a cauldron. The vacuum is no longer a "void" and Faraday would hardly recognize his invention.

In the fifth chapter we look at the most important solutions of the Maxwell equations for the electromagnetic field—those describing light waves, radio waves, and so on. We discuss solutions of the linear wave equation in one, two, and three dimensions, vibrating strings and membranes, sound and light as wave phenomena; superposition, interference, beats, the Doppler effect, and Schrödinger waves in quantum mechanics, all leading to the dominance of linear equations in the physics of the twentieth century. The chapter ends with the discovery of the concept of solitons, waves that behave in some ways like particles, and the growing attention to nonlinear phenomena.

We encountered the special theory of relativity for the first time in Chapter 4; Chapter 6 takes it up in more detail. We introduce the Lorentz transformation, which is the relation of length and time measurements in laboratories that move relative to one another, discuss the slowing of clocks, the Lorentz contraction, and the puzzling twin effect, the different rate at which a traveling twin ages as compared with her stay-at-home sibling. Everything is explained geometrically by means of Minkowski diagrams without any algebraic calculations. We discuss in some detail how the combination of the theory of relativity and our usual notions of causality prohibit the existence of signals and particles that travel faster than light (tachyons).

In Chapter 7 we turn to the question of causality as seen from the point of view of quantum mechanics, and to the philosophical revolution in the notion of reality brought about by the quantum theory. We introduce various quantum interpretations and encounter one of the great debates in Western intellectual history, occasioned by the famous "EPR paper": Does quantum mechanics provide a complete description of reality? *No* said Einstein, *yes* said Bohr, and we learn what they meant by their answers. We discuss Bell's inequality, whose experimental violation demonstrates either the absence of Einsteinian reality or the existence of "spooky action at a distance."

In Chapter 8 we trace the historical development of the concept of an "elementary particle" from the ancient Greeks to the present day and find that there has been a great change. In order to understand contemporary views, we need both the quantum theory and relativity, as discussed in earlier chapters. Not only are there many more kinds of particles than the Greeks dreamed of, but many of the newly discovered particles are unstable. What do we mean when we say that a particle exists if it lives for only 10^{-23} seconds and is far too small to be seen even under a microscope? High-energy physicists usually discover a particle as a "resonance" or a bump on a data curve. Why is it legitimate to interpret such bumps as particles? Particles are divided into two classes, bosons and fermions, and we interpret their different roles. We also question the notion that all the many newly discovered particles are elementary and talk about the invention of classification schemes by means of a smaller number of "fundamental" particles.

Up to this point in the book we have discussed elementary systems of particles and forces or we have reduced large systems to their smaller and more elementary constituents. In Chapter 9 we turn to large-scale phenomena whose explanations are not directly reducible to simpler ones; these are systems for which the whole is truly greater than the sum of its parts. Our understanding of such cooperative effects is based on the quantum theory, and most of the phenomena of this nature occur in the discipline of condensed-matter physics, which deals with matter in bulk. We discuss the general nature of phase transitions and their specific occurrence in ferromagnetism, that is, the behavior of familiar permanent magnets, as well as superfluidity and superconductivity, which, not long ago, captured the public's imagination. Much of the solid matter under discussion in this chapter is arranged in the highly structured form of crystals, whose main characteristic is their symmetry. In the tenth chapter we turn to the consideration of the general importance of symmetry in the formulation of physical theories.

As we had left unexamined in Chapter 8 the basis on which classification schemes for elementary particles by means of more fundamental ones are constructed, in the final chapter we discuss the most important ingredient of this basis, the use of symmetry considerations. We introduce, with examples, the concept of symmetry as used in the arts of various cultures and its mathematical formulation by means of the theory of groups. We then turn to the use of symmetry or invariance in physics. An important mathematical result called Noether's theorem allows us to infer the conservation laws of energy, momentum, and angular momentum from invariance or symmetry properties of isolated physical systems, both in classical physics and in quantum mechanics. Other symmetry considerations in quantum mechanics are used with powerful effect in modern theories of elementary particles and their classifications. Broken symmetry has its own special aesthetic appeal and also plays an important role in particle physics.

A brief epilogue sums up some of the book's conclusions, perhaps the most important of which is that the concepts and theories of science are determined not by facts alone but also, in the words of Galileo, by what is "pleasing to the mind." I hope that our journey through the world of physics will be as pleasing to your mind as it is to mine.

1

SCIENCE, MATHEMATICS, AND IMAGINATION

"Fact, fact, fact!" said the gentleman. And "Fact, fact, fact!" repeated Thomas Gradgrind.

"You are to be in all things regulated and governed," said the gentleman, "by fact. We hope to have, before long, a board of fact, composed of commissioners of fact, who will force the people to be a people of fact, and of nothing but fact. You must discard the word Fancy altogether. You have nothing to do with it."

To many people, the collection of *facts* is the defining characteristic of science, and scientists are thought to be clones of Thomas Gradgrind and his visitor in Dickens's *Hard Times*. Do scientists indeed "discard the word Fancy altogether" and spend their time and efforts dryly deducing laws of nature from observations? Nothing could be farther from the truth. Imagination, passion, and ideas play at least as important a role in the development of science as in any other creative field of human endeavor.

The motives behind our distant ancestors' first attempts to comprehend nature at large, rather than only their immediate surroundings, were both utilitarian and mystical-spiritual. They pursued astronomy in order to make predictions of the seasons for the purpose of better harvests, and also to forecast such awe-inspiring natural phenomena as eclipses of the sun and the moon. For the Babylonians as well as for the builders of Stonehenge, the primary stimuli for primitive astronomy included a need for religious ceremonies.

To study nature in a systematic fashion for its own sake, simply in order to satisfy our urge to understand it, is one of the great legacies of ancient Greek civilization. We may think of the philosopher Democritus of the fifth century BC, who regarded certain hard, solid, invisible, small particles that differed from one another only in shape and arrangements—his *atoms*—as the ultimate constituents of matter, or of Archimedes of the third century BC and his law of the buoyancy of floating bodies. That is not to say, of course, that Greek science did not have any utilitarian or ritualistic motivations. Ptolemy and his followers used astronomy to predict eclipses for religious purposes, Pythagoras approached mathematics with deeply mystical aims, and Archimedes applied his science directly to military purposes. There is, nevertheless, little doubt that the more-than-casual attempt to understand the inner workings of nature, purely for the sake of that understanding, began with the Greeks. The same is true of mathematics. While the beginnings of geometry go back to Babylonian and Egyptian methods of land surveillance, the ancient Greeks were the first both to systematize geometry and to study numbers as such.

Although what we would now call "pure science" and "pure mathematics" originated some twenty-five centuries ago, both languished for a very long period of time, not forgotten but frozen for about 1,700 years. The only exception to this was the significant advancement of algebra by the Arabs. (The word "algorithm" for a systematic mathematical procedure originated from the name of the mathematician and astronomer al-Khwārizmi, and the word "algebra" is a distortion of the first word of the title of his book *Al-jabr wa'l muqābalah,* which means "rearrange," referring to the rearrangement of terms to solve an algebraic equation.) While a number of important technological developments, particularly among the Chinese, occurred during this period of time (many of them later duplicated in Europe), there was essentially no progress in basic science.

Modern Science

Modern science, in the sense in which we now understand that term, began no earlier than the sixteenth century. The credit for its initiation is usually given to Galileo Galilei, born in 1564, and Isaac Newton, born in 1642, the year of Galileo's death. During the last 300 years the rise of science has been truly spectacular. Not only has our knowledge enormously increased, but the number of people who devote a large part of their lives to the advancement of that knowledge has grown impressively. The American Association for the Advancement of Science, for example, which was founded in 1848 with 461 members, now has a membership of over 130,000. It is an often-repeated fact that more than half the scientists who ever lived during the entire history of the world are alive today.

What were the fundamental aims and purposes of the scientists and mathematicians who contributed to this explosive expansion of knowledge? Compare the works of two men of the Italian Renaissance, a century apart, Leonardo Da Vinci and Galileo Galilei. In addition to being a painter, designer, and architect, Leonardo was a most ingenious inventor of technical devices, and he offered the services of his technical imagination and ingenuity to dukes and princes for the enhancement of their military power. But even though medical science undoubtedly benefited from the drawings of his studies of the inner structure of the human body, we do not consider him a scientist. On the other hand, even though duplicates of Galileo's telescope became important tools for navigation and were, at the beginning, used for that purpose more than any other, Galileo did not consider such applications his primary aim, and we regard him as the modern scientist *par excellence*.

The main reason for crediting Galileo and Newton with the origin of modern science is that they based their search for knowledge of nature on observation and experiment and did not believe that such knowledge could be gained by pure thought alone. This was their revolutionary advance over their Greek intellectual predecessors. In addition, their quest was not driven by any desire for useful applications. Though they were neither hostile nor even indifferent to such applications, their basic motivation was not to seek new knowledge for the benefit of society, or to enhance the power of their nation, king, or duke. It was to *understand* the world of nature. There can be no doubt that the urge to understand, to decodify the powerful universe around us had, for some, an aesthetic motivation, and for others a mystical, perhaps even a religious, component. Consider what John Maynard Keynes, the English economist whose hobby it was to collect Isaac Newton's unpublished manuscripts, said about the great scientist:

In the eighteenth century and since Newton came to be thought of as the first and greatest of the modern age of scientists, a rationalist, one who taught us to think on the lines of cold and untinctured reason. I do not see him in this light. I do not think any one who has pored over the contents of the box which he packed up when he finally left Cambridge in 1696, and which, though partly dispersed, have come down to us, can see him like that. Newton was not the first of the age of reason. He was the last of the magicians, the last of the Babylonians and Sumerians, the last great mind which looked out on the visible and intellectual world with the same eyes as those who began to build our intellectual inheritance rather less than 10,000 years ago. Isaac Newton, a posthumous child born on Christmas Day, 1642, was the last wonder-child to whom the Magi could do sincere and appropriate homage.

Pure Science and Social Benefits

The 300 years since the publication of Newton's *Philosophiae naturalis principia mathematica** have seen not only a great growth in our understanding of nature but also a large advance in technical knowledge in all areas of life. In those regions of the world that have had direct access to this technical knowledge, the standard of living, health, and life span of the vast majority of the population have greatly increased. Few would deny a connection between these two developments. In fact, almost everyone credits the advances in science with the social benefits of the past three centuries.** The result of the correctly perceived correlation between advances in science and social benefits is that many people think one can enhance this correlation by supporting and encouraging primarily those scientists whose motivation is not the disinterested, pure desire to understand nature but the wish to benefit mankind, those scientists, therefore, who do research that is relatively close to applicability. Such a policy would be short-sighted and, in the long run, counter-productive.

It is generally acknowledged that a man like Thomas Edison, who was not a scientist but an inventor, made many important contribu-

*The international language of scientific and other learned discourse in Western culture at the time was still Latin, much as today it is English.

**It is undoubtedly true, and sometimes held against science to an extent that outweighs acknowledgment of its benefits, that there have also been great social costs, such as environmental degradation, the increased destructiveness of wars, and the pains caused by social change.

tions that were extremely useful to society at large. But such technical progress occurs only rarely and slowly without the existence of a large amount of basic understanding of the world. It is surely no accident that the number of technological advances in China, and in Europe before the Renaissance, was quite small when compared with that number during and after the eighteenth century in Europe. Most prominent European scientists and mathematicians did not completely ignore applications of their work. Indeed, the mathematicians Pierre-Simon de Laplace and Joseph-Louis Lagrange, in much of their important work in celestial mechanics, were motivated to some extent by the desire to determine if the solar system was stable or whether there was some danger that humanity's life might be cut short by a celestial catastrophe. Very few scientists or mathematicians, especially in the nineteenth century, took the attitude of the influential English mathematician Geoffrey H. Hardy, who proclaimed himself proud never to have worked in an area of mathematics that had the slightest chance of ever having any use or application. (He turned out to have been wrong in this. The theory of numbers, which was his primary field of research and in which he made many important contributions, was later found to be extremely useful to cryptography.) In fact, the kind of dichotomy between pure and applied science, or between pure and applied mathematics, that we have today did not exist during the nineteenth century. Nevertheless, the great scientists Galileo, Harvey, Lagrange, Laplace, Darwin, Boltzmann, Gauss, Faraday, Maxwell, Gibbs, Planck, Morgan, Einstein, the Curies, Bohr, Rutherford, Muller, Heisenberg, Schrödinger, Dirac, Fermi (most of whom we shall encounter again in this book), and many others too numerous to mention, were not motivated primarily by the desire to find new ways to be useful to the world. *They wanted to understand what made nature tick.* As Albert Einstein put it, "I want to know how God created this world . . . I want to know his thoughts," and "What really interests me is whether God had any choice in the creation of the world."

The science of medicine clearly forms an exception to this entire argument. The motivation of the scientist here is and has to be largely to benefit humanity. However, even in this field it is well to remember the words of Hippocrates hewn in stone on medical science buildings: *The nature of the body is the beginning of medical science.* Today, of course, we include in that nature of the body its microscopic constituents. Among biologists, Louis Pasteur was a notable exception.

Some of his greatest discoveries were motivated directly by the desire to solve urgent agricultural or industrial problems.

Beauty as Motivating Force

In many instances the driving motivation of theoretical scientists and mathematicians is so far removed from ideas of applicability that it is closer to that of the artist. While for some there is a mystical awe of nature that approaches a religious feeling, for many others it is an admiration of beauty. "What I remember most clearly," the astrophysicist Hermann Bondi wrote, "was that when I put down a suggestion that seemed to me cogent and reasonable, Einstein did not in the least contest this, but he only said, 'Oh, how ugly' . . . He was quite convinced that beauty was a guiding principle in the search for important results in theoretical physics."

To be sure, just as some other greatly admired kinds of beauty require a highly educated taste to be fully appreciated, so the beauty of a mathematical structure, of an equation, or of a physical theory can be appreciated only after the beholder has acquired the necessary kind of training and knowledge. We must, in other words, learn the appropriate language in order to understand what is being communicated. In the physical sciences this language is always that of mathematics.

There is, however, one essential difference between the aesthetic motivation of an artist and that of a scientist or mathematician: while the artist is subject to no other authority, the scientist has to bow before the final arbiter of "truth" as revealed by experiment or observation. Similarly for a mathematician: no matter how beautiful a purported theorem is, it has to be logically correct in order to be accepted as a theorem. Nevertheless, aesthetics plays an important role both in the initial plausibility of an announced result and in the value it is accorded after its proof has been accepted. Even though the great Indian mathematician Srinivasa Ramanujan, who grew up without the benefit of extensive education in traditional mathematics, announced many astonishing mathematical propositions without proof, his English mentor G. H. Hardy and other mathematicians were almost immediately convinced of their truth in part because of their beauty. This, of course, did not obviate the need for actually proving them before they could be accepted as theorems (which sometimes turned out to be difficult), but although a few were found to be incorrect, beauty contributed to the *prima facie* case for truth. Not only do mathematicians accept beautiful results more readily than ugly ones, but they also greatly admire beautiful and elegant proofs and put a lower value on clumsy and unnecessarily complicated ones.

Are Theories Deduced from Experiments?

When Isaac Newton made his famous disclaimer, "I do not make hypotheses," he meant that he did not indulge in speculation, as was sometimes customary among philosophers and scientists at the time; his gravitational theory was grounded in observed evidence. Nevertheless, such evidence does not lead to a general theory by a purely logical process. What is required, as Einstein put it, is "intuition supported by rapport with experience." Indeed, the axioms that form the basis of fundamental physical theories, Einstein averred somewhat hyperbolically in his Spencer Lecture in 1933, are "free inventions of the human intellect." He did not mean to deny that the "free inventions" must be finally anchored in empirical observations, but they are not determined by them. "Experience may suggest the appropriate mathematical concepts," Einstein continued, "but they most certainly cannot be deduced from it. Experience remains, of course, the sole criterion of physical utility of a mathematical construction. But the creative principle resides in mathematics."

There is an enormous gap between falling apples, rolling balls, and planetary orbits on the one hand, and the laws of motion and the law of universal gravitational attraction on the other; and "there is no logical bridge from experience to the basic principles of theory" (Einstein). In addition to the original inference that leads from "rapport with experience" to a scientific theory, there is the further constraint that predictions of the theory must be verified by subsequent observation. "The process by which we come to form a hypothesis is not illogical but non–logical, i.e., outside logic," wrote the biologist Peter Medawar, "but once we have formed an opinion we can expose it to criticism, usually by experimentation."

Between a scientist's soaring imagination and the need to discipline that free flight of thought by critical comparison with empirical fact there is a continual tension. "Let the imagination go, guiding it by judgment and principle," advised Michael Faraday, "but holding it in and directing it by *experiment*." It is this second aspect of the scientist's creation that distinguishes him from the artist, who is not subject to such constraint. And it is this aspect, Ernest Rutherford emphasized, which assures us that "the physicists have . . . some justification for the faith that they are building on the solid rock of fact, and not, as we are often so solemnly warned by some of our scientific brethren, on the shifting sands of imaginative hypothesis."

The need for firm grounding in empirical facts lies at the basis of physicists' insistence on *quantititative* agreements between theoretical predictions and empirical observations; we do not really understand a natural phenomenon unless we can predict its occurrence under specified circumstances in quantitative terms. This emphasis on quantitative predictions as a test of real understanding is sometimes puzzling and alienating to nonscientists. When a historian claims to understand a course of events, the implication is not necessarily that this understanding, had it been available beforehand, would have allowed anyone to predict with certainty precisely what happened. Scientists, however, are suspicious of allegations that someone has really understood a physical process and yet is unable to make precise predictions based on that understanding—they regard such a claim as mushy.

The basic reason for the emphasis on quantitative predictions is not the use of mathematical language per se; not all the mathematics employed by scientists is intrinsically quantitive in the sense that it deals with numbers. The defining characteristic of mathematics is *precision of thought* rather than numerical formulation. In many instances a theory is temporarily accepted long before methods have been developed that would allow a quantitive confrontation with empirical data, either because the equations in the theory are too difficult to solve, or because the predictions lie in an area that is not yet technically accessible to experimentation, or because no experiment is possible and the needed observations of the natural world have not yet been made. The theory may be accepted in the interim because its consequences appear to agree with the phenomena *qualitatively*. In other recent cases, physicists have developed mathematical models that are known not to correspond directly to reality; they serve in part as training grounds for mathematical attacks on more realistic but more complicated theories, and in part as metaphors for the real thing, which no one as yet knows how to formulate. There is no doubt, however, that ultimately the proof of the pudding is in the eating, and the eating of a physical theory consists in *quantitative* confrontation with empirical measurements.

SCIENCE AND IMAGINATION

The Role of Mathematics

If imagination, disciplined by experimentation and observation, is the indispensible tool of the physicist, the framework within which this imagination operates in most instances is mathematics. "A physicist builds theories with mathematical materials," correctly asserts the physicist Freeman Dyson,

and the physicist's art is to choose his materials and build with them an image of nature, knowing only vaguely and intuitively rather than rationally whether or not the materials are appropriate to his purpose. After the design of the theory is complete, rational criticism and experimental test will show if it is scientifically sound. In the process of theory building, mathematical intuition is indispensable.

Thus the most fruitful function of mathematics for physicists is that of a general framework of thought, the most powerful method of abstraction with which to analyze nature and to formulate our description of it. An important part of this function is the *language* of mathematics.

The use of mathematical language, of course, makes the scientist hard to understand for nonscientists. But it should not be thought that elimination of the mathematical language would make the science more comprehensible to the public at little cost to the scientists, for it is not the language alone that plays a role as a form of technical jargon analogous to the jargon employed by many other specialized disciplines. The results and the abstract ideas of mathematics are even more important to the development of physics than the language in which they are expressed. Indeed, it is truly remarkable to what extent mathematical ideas that were formulated with no intention of application to reality turned out to be useful in the construction of physical theories, despite the fact that mathematicians are guided primarily by motives of the power of abstraction, of aesthetic appeal, or of economy of thought. The "purer" mathematicians are, the farther removed they are from any worry about whether their ideas have contact with reality or application.

At the same time it is also true that many important fields of mathematics received their first stimulus from physical applications or ideas about the real world. The best-known example is the invention of the calculus by Newton, and we shall encounter many other examples in later chapters. But even in branches of mathematics whose value to physics was known, many of the powerful advances were made by pure mathematicians without any specific regard to their application, and these advances then often turned out to have great utility, after all, in the formulation of our thoughts about nature.

There is a real enigma at the heart of the relation between pure mathematics and its applicability in science, which Eugene Wigner discussed at some length in an essay aptly entitled "The unreasonable effectiveness of mathematics in the natural sciences." Not only does every branch of physics employ mathematics to various degrees of abstraction (and I do not mean just that numerical calculations are needed for comparison of the theory with experiments), but it is also true that if there is any field of mathematics that has not as yet found applicability to science, it is a fair bet that it will do so in the future. Conversely, physicists have invented mathematical notions that were initially scorned by mathematicians as incomprehensible, and yet they later blossomed into respected areas of mathematics under the guidance of experts.

Some Mathematical Concepts Used in Science

An example of a purely mathematical concept that became extremely useful in science is the extension of the number system to include the "imaginary" number $i = \sqrt{-1}$ and its multiples. Such numbers first surfaced in the sixteenth century as "impossible" quantities that appeared in solutions of algebraic equations. (For example, the solutions of the equation $x^2 + 1 = 0$ are $x = \pm i$. Before the invention or discovery of imaginary numbers, the equation was considered to have no solution.) They were first clearly included among the acceptable solutions of such equations by the Flemish mathematician Albert Girard. It was the philosopher and mathematician René Descartes who called these numbers "imaginary," as we do today, but it took the "prince of mathematicians," Carl Friedrich Gauss, to make them entirely respectable. (Even the concept of *negative* numbers as solutions of algebraic equations did not arise until the fifteenth century and was also difficult to accept at the beginning.) Numbers that are sums of "real" and "imaginary" numbers, as in $3 + 5i$, are now called "complex"; 3 is the "real part" and 5 is the "imaginary part" of $3 + 5i$.

While the concept of imaginary numbers arose in algebra, it has become an integral element of other parts of mathematics, and it was gradually recognized as very useful in many areas of science. The advent of quantum mechanics in the 1920s, which we will discuss in more detail in later chapters, enormously bolstered the concept; the complex number system is a basic ingredient of that theory, and its formulation would be hard to envisage without such quantities.

Non-Euclidean geometry is another example of a branch of pure mathematics that acquired great physical usefulness. Euclid's fifth postulate, the parallel postulate, which states that through a given point there is exactly one straight line parallel to a given line, had for a long time played a special role among the axioms of geometry. It was suspected by some not to be independent of Euclid's other axioms, and mathematicans had tried unsuccessfully many times to show that the parallel postulate follows from the remaining four; others considered its "truth" more or less self-evident. The influential philosopher Immanuel Kant regarded Euclidean geometry, including the fifth postulate, as one of the truths about nature that could be arrived at by pure thought.

About 150 years ago, less than fifty years after Kant's death, the mathematicians Nikolai Lobachevsky, Farkas Bolyai, and Georg Friedrich Bernard Riemann constructed geometries all of whose axioms are identical to Euclid's, except the fifth. Lobachevsky and Bolyai substituted the postulate that to any given line and through any given point there is more than one parallel, and Riemann constructed a geometry in which he assumed that there is none. These geometries were shown to be fully consistent, and hence their existence proved that Euclid's fifth axiom is independent of all the others. While in Euclidean geometry the sum of the angles in a triangle is 180°, in Lobachevky's "hyperbolic" geometry the sum of the angles in a triangle is less than 180°, in Riemann's "elliptic" one it is greater than 180°. In both, the sum of the angles depends on the size of the triangle.

One can easily construct models of such geometries in two spatial dimensions. If we call a great circle on a sphere a straight line (because it is the shortest path between two points on that sphere), then two-dimensional geometry on the surface of a sphere is Riemannian. The geometry on the surface of the earth is a good example. Form a large triangle consisting of a line running straight north-to-south through New York, a second line running north-to-south through Los Angeles, and a third line along the equator. The two interior angles at the corners on the equator are each 90°. Therefore the sum of the interior angles of this triangle exceeds 180° by the angle at the third vertex at the north pole, which equals the difference in longitude of the two cities, New York and Los Angeles. There are no parallel lines in this geometry, since all great circles intersect at two points. An analogous geometry on a "pseudosphere," an infinitely long trumpet-shaped surface, is hyperbolic.

In three dimensions these were purely mathematical constructions, done primarily for the purpose of showing the independence of Euclid's fifth axiom. The thought of applicability to the real world could not have been farther from the minds of the inventors (although Gauss already speculated about the possible reality of hyperbolic geometry). But in 1916 Einstein invented a new theory of gravitation, the *general theory of relativity,* one of whose foundation stones is the fact that all bodies are subject to the same gravitational acceleration, irrespective of their mass. This was of course known in the Newtonian theory, but there it had been accepted more or less as an accident without special significance. By contrast, the fact that so long as an object is subject to gravitational forces only its mass cancels in the equations of motion became central in Einstein's theory. Since, as a result, all bodies behave in the same way, Einstein was able to describe the entire motion in purely geometrical terms and to envisage gravitation directly as a property of space; the "natural" geometry of the space cannot then be Euclidean: we say that "space is curved."

To be more concrete, let us call the path of a light ray a straight line. This is a natural thing to do if we consider that the most common way of testing whether an edge is straight is to hold it to your eye and sight along it. The geometry of such light rays in Einstein's general theory of relativity turns out to be Riemannian! A large triangle made up of light rays includes three angles that add up to more than 180°; how much more depends on the size of the triangle. For triangles that are the size of everyday objects, the deviation from 180° is imperceptibly small, becoming measurably large only at astronomical distances.

So what started out as an abstract mathematical exercise ended up having direct physical application. What is more, the whole geometrical approach to gravitational theory has had a deep effect on the thinking of some theoretical physicists. We imagine strong forces as "warping space," and inside superdense stars this warpage can be enormous. The motion of comets, asteroids, or planets is, from this point of view, as "natural" and "straight" as that of a free body in Newtonian mechanics, subject to no forces whatever. They are simply

SCIENCE AND IMAGINATION

moving along "geodesics," that is, paths which, in the appropriate geometry, connect two points by the shortest route.

The Mystery of the Efficacy of Mathematics

The intimate relation between mathematics and the natural sciences remains largely a mystery. No doubt in some cases physicists who are aware of certain abstract branches of mathematics are stimulated to follow ideas that have been worked out in those areas, and they are led to express their thoughts in that language. But in other instances the physicists are, in fact, ignorant of what has been done by mathematicians, and they begin to formulate their ideas in a manner that is subsequently found to have been known to mathematicians without any thought to application.

Possibly the explanation of the uncanny correlation between the abstract ideas of mathematicians and our description of nature is simply the fact that mathematics is no more than a very powerful codification of logical short cuts; scientists need all the logical thinking tools available for the formulation of what their imagination tells them to be the nature of the universe, and for drawing as many conclusions from their ideas as possible. Mathematics serves as an abstract and far-reaching organizing principle, without which our ideas would remain inchoate. "The mathematics enables [us] to imagine more," wrote Freeman Dyson, "than [we] can clearly think."

If the basic ideas of physical theories and the mathematical concepts in terms of which we explain the workings of nature, though tied down by experimental facts, are "free inventions of the human intellect," then we may well ask, could entirely different theories, possibly using quite different mathematics, perform the same function with equal or perhaps even greater efficacy? This is a very controversial question among scientists. If physical theories and mathematical theorems are, indeed, human inventions, then either alternatives of comparable merit must exist, or else a remarkable kind of harmony between the workings of the human mind and nature as a whole, postulated by some philosophers, has to exist.

In later chapters we will encounter specific examples of the lack of uniqueness in the correspondence between natural phenomena and their description by physical theories: there usually is an element of simplification involved. In order to arrive at a manageable theory that can be formulated in terms of relatively simple mathematics, we always have to make certain simplifying assumptions that are known to be violated by the actual phenomena. To describe the real observations, physicists rely, when necessary, on needed "corrections" whose nature, of course, is not arbitrary but based on further theory. The process of simplification that leads to a theory is not unique, and is, in each instance, influenced either by the existing mathematical tools or by the scientist's imaginative invention of new ones. Conceivably, therefore, other theoretical approaches might simplify the phenomena in different ways and arrive at equally valid descriptions that have little in common with the theories we now accept.

But if theories are "free inventions," are they really telling us anything about the world, or are they mere reflections of the structure of our own mind? The fact that these theories serve to predict future observations and enable us to construct elaborate working technologies certainly proves that they say *something* about nature, even if that something is not uniquely expressible. There is no inherent contradiction between believing that scientific theories are largely products of the human imagination and also insisting that they tell us *truths* about the world.

A stimulating way of pondering such questions is to imagine a visit of alien beings from another planet. Would their physical theories and their mathematics necessarily be the same as ours? Or at least could their theories, after certain rearrangements and superficial shifts of the order of ideas, necessarily be put into one-to-one correspondence with ours? In mathematics it is very hard to imagine that an alien culture could have developed very highly without the use at least of arithmetic; we can't be so sure about other parts of mathematics. That is why some mathematicians say, "God made the integers, but everything else is man-made." Similarly, there are surely some basic ideas of science, such as the periodic table of the elements, that have to have a counterpart in any highly developed civilization, but I am not convinced that the same is true for all other parts of our scientific edifice. However, I do not think that a definitive answer to this question, and therefore, at the same time, a solution to the problem of the correspondence between human imagination and the external world, is possible without the observational evidence of a visitor from outer space.

In the meantime we are free to admire the conceptual structures that have been erected, which are as imposing as the spacious vaults and soaring spires of medieval cathedrals, and were built with as little utilitarian purpose. Let us now turn to an inspection of the architecture and the various chapels of the cathedral.

SCIENCE AND IMAGINATION

2

CHAOS AND THE GHOST
OF LAPLACE

Given for one instant an intelligence which could comprehend all the forces by which nature is animated and the respective situations of the beings who compose it—an intelligence sufficiently vast to submit these data to analysis—it would embrace in the same formula the movements of the greatest bodies and those of the lightest atom; for it, nothing would be uncertain and the future, as the past, would be present to its eyes.

So wrote the great French mathematician Pierre-Simon de Laplace in 1795, and his grandiose claim exerted a far-reaching influence on the philosophical outlook of most of nineteenth-century Europe and America. The universe was regarded by scientists and many philosophers of that time as a clockwork; if only we had sufficient information and were smart enough, we could know and predict its entire future with unlimited precision. Since in such a mechanistic world there was no room for God and divine intervention, the chasm that separated science and religion, which already existed with painful consequences in the time of Galileo, became unbridgeable.

What I want to explore in this chapter, our first foray into a large set of fundamental concepts in the description of nature, is the basis of Laplace's claim. We shall find that, even in the context within which he worked and made the statement,* it is of debatable validity. We will be led from a world of classical order to a cauldron of chaos, and along the way there will be vistas the memory of which will serve us in other contexts. In the course of our exploration, we will have to introduce a certain minimal amount of mathematical apparatus, which will also come in handy in other chapters. The trip begins in the seventeenth century.

The principles of classical mechanics, upon which Laplace's claim was based, were completely formulated by Isaac Newton's laws of motion:

1. A body remains in a state of rest or uniform motion unless acted upon by a force.
2. A force acting on a body causes it to accelerate, and this acceleration is proportional to the strength of the force.
3. To every force acting on a body there is an equal and opposite force of reaction.

These laws, which incorporated Galileo's demonstration that the force of gravity causes a specific acceleration, constituted a complete break with the Aristotelian notion that even to maintain a body in a state of uniform motion required the action of a force. Indeed, even today, that view of Aristotle has such a powerful hold on the intuition of many people that they find Newton's equations hard to accept.

In mathematical terms we would say that, according to Newton's second law, a force acting on an object of mass m will cause it to accelerate by $\vec{a} = \vec{F} / m$. The arrows are meant to indicate that these quantities have directions as well as magnitudes, which needs a bit of explanation.

Many quantities in physics are characterized not only by a number that specifies their magnitude but also by a direction. A force is an example of such a quantity, and so is the velocity of an object. Such quantities are called *vectors,* and they are usually shown by an arrow above them, such as $\vec{v}$. For more on vectors see the box on the next page.

---

*As we shall see in Chapter 7, the quantum theory invalidates Laplace's dream in any event, but we shall confine ourselves here entirely to classical mechanics. In spite of the fact that in a certain sense quantum mechanics supersedes classical mechanics, for macroscopic objects classical mechanics is still a valid theory.

In the physical space in which we live, specifying three numbers in a given coordinate system designates the location of a point. For example, in a room two of whose walls face west and north, we may use the northwest corner on the floor as the "origin" and then specify the location of a point by giving its distance from the north wall, its distance from the west wall, and its height from the floor. These three numbers, usually designated by x, y, and z, are called its *Cartesian coordinates* after René Descartes, or simply its coordinates. If any of these numbers is negative, the point lies outside the room; if z is negative, the point lies below the floor, and so on. In the same manner, one may specify any vector $\vec{r}$ by drawing an arrow whose back end is located at the coordinate origin and whose head has the coordinates r_x, r_y, r_z. One then says that the vector has the "components" (r_x, r_y, r_z). (See Figure 1.) Thus, to specify a vector, one may give its magnitude and direction in some fashion, or one may give its three components. Any given set of three numbers may be regarded as specifying the components of a vector, or the location of a point in three-dimensional space. If we draw a line from a fixed point, which we call the "origin,"

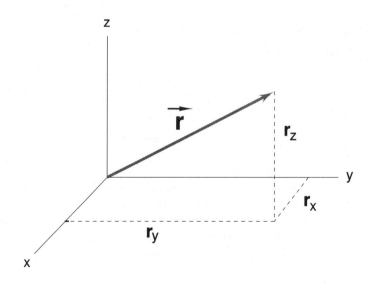

FIGURE 1. A vector $\vec{r}$ and its components, r_x, r_y, and r_z.

to the position of an object, we may regard the position too as a vector.

This idea may be generalized to any set of n numbers. A point that requires n numbers to designate its location is said to lie in an n-dimensional space. If one number suffices, the point simply lies on a line, and the space is one-dimensional; if two numbers are required, the point lies in a plane, which we call two-dimensional. If more than three numbers are needed, we cannot form an image of the space in our minds, and there is no need to try, but the notion of such spaces, as we shall see, is extremely useful.

The Newtonian Equations of Motion

If a physical system consists of several objects, there is a Newtonian equation of motion for each of them, and if these objects exert forces upon one another, the force $\vec{F}_1$ on the first may depend not only upon the position of that body but also upon the positions of all the others; similarly for each of the other bodies. In the planetary system around the sun, for example, the forces exerted on the earth are the gravitational force of attraction pulling in the direction of the position of the sun, plus the pull exerted by the moon, plus the much weaker attractions from the other planets. As you can see already, this may become complicated, and for systems consisting of many bodies, it becomes very complicated indeed.

If the forces between all the particles* or bodies in a mechanical system are known in their dependence upon the distance, and all other forces on the particles are known, then the Newtonian equations of motion for all the constituents of the system can be written down. They are *differential equations* and employ the language of the calculus.

In his *anno mirabilis*, 1665, immediately following his A.B. degree from Trinity College, Cambridge, Isaac Newton invented (or, as some mathematicians would prefer to say, discovered) the differential calculus, specifically with application to physics in mind. There was, however, a second, independent inventor or discoverer a number of

---

*We shall use the word *particle* here simply to denote a small, pointlike object, without meaning to imply anything about its divisibility. In Chapter 8 the word will be used differently.

years after Newton's original work: the German philosopher and mathematician Gottfried Wilhelm Leibniz, arguably the last universal polymath. The fact that Leibniz's idea was published in 1684, before Newton's work appeared in print in 1687, gave rise to ugly priority squabbles and bad feelings between followers of the two men, which had the unfortunate after-effect of isolating British mathematicians from their continental colleagues for many years after Newton's death. Today Newton's priority of the idea and development of the calculus is almost universally acknowledged, though the notation generally used originated with Leibniz.

A string of earlier contributions by, among others, Galileo's student Buonaventura Cavalieri, the French mathematicians Pierre de Fermat, whom Laplace called "the true inventor of the differential calculus," and Blaise Pascal had led in the direction of and prepared the ground for the work of Newton and Leibniz. Their "infinitesimal calculus" was the beginning of the development of a branch of mathematics now called *analysis,* which marked a completely new departure from the mathematical ideas and concepts that had come down from the ancient Greeks and the Arabs. Though other parts of mathematics played important roles as well, it is this field of mathematics above all others that made modern science possible. (For more on functions, differential calculus, and differential equations, see the boxes on the next page and on pages 31 and 36.)

As a significant general characterization, which tells you much about the way our concepts of nature are formulated, we may say that a differential equation in physics, such as Newton's equations of motion and others we shall encounter later, is an equation for physical quantities at nearby points in space or time; the equation does not directly relate to one another quantities at distant points, but only at points in each other's infinitesimal neighborhood.

Because the acceleration $\vec{a}$ is the derivative of the velocity regarded as a function of the time, and the velocity in turn is the derivative of the position $\vec{q}$ of a particle, the acceleration $\vec{a}$ is the second derivative of the position $\vec{q}$. The Newtonian equations of motion are therefore differential equations of the second order; since there is but one independent variable, time, they are called "ordinary differential equations." By a solution, we mean that all the positions or coordinates $\vec{q}_1$ $\vec{q}_2$, ... of the particles are known for all times: $\vec{q}_1(t)$, $\vec{q}_2(t)$, ... (The subscript enumerates the particles; $\vec{q}_1$ is the position of the first particle, $\vec{q}_2$ that of the second, and so on; these are the dependent variables.)

FUNCTIONS

To understand the Newtonian equations of motion and the differential calculus upon which they depend, we must start with the concept of a *function*. Let us denote by x a variable quantity (mathematically called simply a "variable") that can take on any number as its value. Another variable y is called a *function of x*, written $y = f(x)$, if, for any specified value of x, y has a unique specified value. In this relationship x is called the *independent variable* and y is called the *dependent variable*, because x may be chosen freely and independently, while y depends on x. For example, if $y = 5x$, then for $x = 2.5$, y has the value 12.5; if $y = ax$, then for any given value of x, y has the value obtained by multiplying x by the specified number a. If $y = x^2$, the value of y is obtained by squaring the value of x, that is, by multiplying x by itself. More generally, a function may be given in an algebraic form that allows us to calculate y whenever x is given. In the expression $f(x)$ the variable x is called the *argument* of the function f.

A function may be given by means of a graph, as in Figure 2. One starts with two base lines, one horizontal (the x-axis) and the other vertical (the y-axis); both have a specified scale, which means that to every number x there corresponds a point on the x-axis, and to every number y there corresponds a point on the y-axis, as indicated in the figure. We determine the value y_1 of y

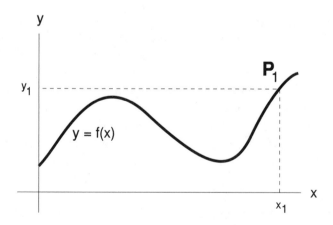

FIGURE 2. The plot of a function $y = f(x)$.

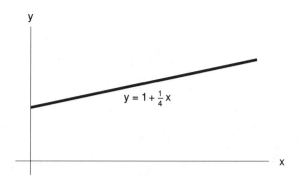

FIGURE 3. The plot of a linear function, a straight line.

that is assigned by the function to a given value x_1 of x by drawing a vertical line from the point x_1 on the x-axis that corresponds to the given value of x upward until it hits the curve, and from that point a horizontal line to the y-axis. The value y_1 of y that corresponds to the point of intersection on the y-axis is the value that the function assigns to the given value x_1 of x. In this manner the curve expresses the functional relationship between the variables x and y.

Such a curve may be given by observational data, or it may be given by an equation. An example of the first would be a plot of the temperature at a certain location and time as a function of the altitude. In that case x would be the altitude above the ground, say, in feet, and y would be the temperature, say, in degrees Fahrenheit. An example of the second would be a plot of the equation $y = 1 + \frac{1}{4}x$, which is a straight line, as in Figure 3, that intersects the y-axis at the point 1 and has a slope of $\frac{1}{4}$ upward. A function whose plot is a straight line is called a *linear function*.

In many cases the dependent variable may depend on more than one independent variable. For example, the air temperature at a certain location depends not only on the altitude above the ground but also on the time; in June it will not be the same as in January, at noon not the same as at midnight.

If the two independent variables are denoted by x and y, and the dependent variable by z, we write the functional dependence $z = f(x,y)$. To exhibit such a functional relation graphically requires a plot in three dimensions, with two mutually perpendicular horizontal axes for the variables x and y, say, with the

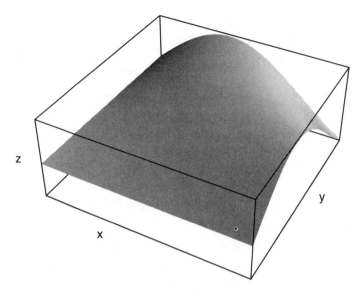

FIGURE 4. The plot of a function $z = f(x,y)$ of two independent variables is a surface in three dimensions.

x-axis running east-west and the y-axis running north-south, and a vertical axis for the variable z. The plot of the function is now a surface S (see Figure 4).

In order to determine the value of z for given values of x and y, we first find the point P in the horizontal base-plane that corresponds to the given values of x and y by drawing a line running north-south through the point on the x-axis that corresponds to the given value of x and a line running east-west through the point on the y-axis that corresponds to the given value of y. The point P is the intersection of these two lines. The next step is to draw a vertical line through P and find its intersection I with the surface S. The intersection of the horizontal line from I to the z-axis with that z-axis determines the value of z that is assigned by the function to the pair of independent variables x and y.

The primary purpose of the differential calculus is to handle the change that a function undergoes as its argument is continuously altered, or, to put it another way, the calculus allows us to find the rate at which a given function changes when its argument is either increased or decreased. In order to describe such a rate of change we must compare the values of the function at nearby points of its argument; the differential calculus is designed to accomplish that in a systematic fashion.

Suppose the speedometer of your car is broken and you want to measure its speed as you drive along. How would you go about doing that? Your first approach would very sensibly be to look for two markers along the road whose distance you know or have determined beforehand, and to measure the time the car took to cover the distance between them; or vice versa, if you are driving along a road with lots of distance markers, you drive for a predetermined time and measure the distance the car covered during that time. In any case, the ratio of these two numbers, the distance divided by the time, would tell you the *average* speed of the car during this interval. If your speed was constant, that is a satisfactory answer. However, if you are not sure that you drove at exactly constant speed and you want a precise answer, what you clearly have to do is to reduce the time interval and the distance covered; the shorter these are, the safer it is to say that the ratio you have calculated was the exact speed of the car at that time, because the average speed during a short enough time interval coincides with what you would regard as the instantaneous speed at that time.

We may plot the distance your car covers as a function of the time, $D = f(t)$, as in Figure 5. If the time interval $t_2 - t_1$ is small enough (it is then called *infinitesimal*), the curve of the plot from P to Q may be regarded as essentially straight, and we may say that the speed of your car at the time half-way between t_1 and t_2 is given by the ratio $(D_2 - D_1)/(t_2 - t_1)$; the speed is the *slope* of the curve at the point P. (Remember that the slope of a highway in the mountains is given by, say, 1/15, which means that for every fifteen feet in horizontal distance, the increase in altitude is one foot.) The shorter we make the time interval $t_2 - t_1$, the more exact these statements become. The ratio

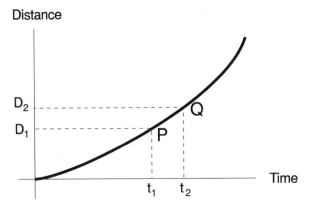

FIGURE 5. The derivative of a function at a point is the slope of its plot there.

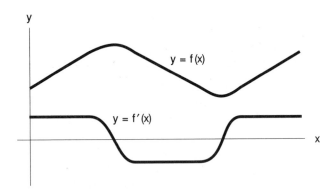

FIGURE 6. Plots of a function $y = f(x)$ and of its derivative function $y = f'(x)$.

$(D_2 - D_1)/(t_2 - t_1) = [f(t_2) - f(t_1)]/(t_2 - t_1)$, that is, the slope of the curve $D = f(t)$, is then called the *derivative* of the function $f(t)$ at the midpoint between t_1 and t_2. The speed of your car as a function of the time is the derivative of the distance it covers as a function of the time; it is the *rate at which the distance D traversed by the car changes with time*. This function, the derivative of $f(t)$, is denoted by either $f_t(t)$ or $f'(t)$. For a given plot of any function $f(x)$, we may plot the corresponding derivative function $f_x(x)$. (See Figure 6.)

To take some other examples of derivatives, if f describes the amount of gas, in gallons, used by your car as a function of the number of miles driven, its derivative is the rate at which it uses gas, that is, the number of gallons per mile. If f is the number of miles driven as a function of the number of gallons of gas used since the last fillup, its derivative is the rate at which the car can cover distances as a function of its gas consumption, the number of miles per gallons of gas. If f measures your weight as a function of the number of calories you consume per day, its derivative is the rate at which your weight increases with your calorie consumption.

Notice that when D_2 is greater than D_1 the derivative is positive and the function $D = f(t)$ grows, or increases, with increasing t; your car is moving in the forward direction. On the other hand, when D_2 is smaller than D_1, $D_2 - D_1$ is a negative number and the ratio $(D_2 - D_1)/(t_2 - t_1)$ is negative; your car is moving backwards. The slope of the curve is then negative and the function $D = f(t)$ decreases with increasing t.

Whenever a function $f(x)$ has a maximum or a minimum at some point x_1, the slope of its plot vanishes there; hence $f'(x_1) = 0$. (The time at which you have arrived at your destination, that is, at which your car has covered its largest distance, is a time at which it has stopped, that is, its speed is zero.) One of the many uses of the derivative function $f'(x)$, in fact, is to search for its zeros and thereby to find the maxima and minima of $f(x)$. This is one instance of a general observation: if we know not only $y = f(x)$ for some value x_1 of the variable x, but also $f'(x_1)$ at that point, then we know something about the behavior of the function in the vicinity of x_1. From the way the derivative was defined as the ratio $f'(x_1) = [f(x_1 + c) - f(x_1)]/c$ for a small increment c, it follows by multiplication by c that the value of $y = f(x)$ for any value of x near x_1, that is, at $x_1 + c$ for any small c, is approximately given by $y \simeq f(x_1) + cf'(x_1)$; the smaller the value of c, the better the approximation. (The sign $\simeq$ is used to denote "approximately equal.") If you want to know how far you have traveled during a short period of time T, you get approximately the right answer if you simply multiply T by the speed of the car at the beginning of T without worrying about its possible variation in speed during T.

Suppose now that we have a given function $f(x)$ and its derivative function $f'(x)$. We may then repeat the procedure and form the derivative of $f'(x)$. This new function is called the *second derivative* and is denoted by $f''(x)$ or by $f_{xx}(x)$. If $f(t)$ is the distance traveled by your car and $f'(t)$ is the speed, the second derivative of f is its *acceleration;* it is the rate at which the car's speed increases with time. If the acceleration a is negative, which we usually call *deceleration,* the speed is decreasing. To take some other examples, if f is the total number of gallons of gas used by your car as a function of its mileage, the derivative of f is the amount of gas used per mile, and its second derivative is the rate at which the gas consumption per mile increases with the mileage on your car. If f gives your weight as a function of your daily calorie consumption, f' is the rate at which you gain weight with consumption, and f'' is the rate at which that weight gain varies. The successive calculation of derivatives may be repeated any number of times, but we shall have no need to make use of that.

In order to solve the differential equations of motion we must be given not only the forces among the particles (and thus the differential equation) but also the initial positions and velocities of each of them, as we shall see in the little boxed excursion on page 36. In other words, the equations of motion for given particles with given forces acting on them have infinitely many solutions, each depending on the positions and velocities of all the particles at the initial instant. The initial conditions are not supplied by the laws of physics but by circumstances, by the world, or by history; they are *contingent.*

The Hamiltonian Equations

The nineteenth-century Irish mathematician William Rowan Hamilton, who began to study Newton at the age of twelve, took a great step forward toward the possibility of solving the complicated Newtonian equations for physical systems consisting of any number of particles. His contributions led to a certain "geometrical" way of viewing the development of a mechanical system, which will facilitate our grasp of the nature of Laplace's claim.

Before we can understand Hamilton's formulation, we have to begin with the concept of *work* or *energy,* a concept that preoccupied much of nineteenth-century science, as we shall see again in the next chapter. Here we are going to deal with it only in the context of

34

mechanics, later more generally. What do we mean by saying that carrying a sack of sand up three flights of stairs is hard work, that is, it requires the expenditure of much energy? We mean that we have to exert an upward force equal to the weight of the sack (to prevent it from falling) and we have to do it for an upward distance equal to the height of three stories: the work is equal to the exerted force multiplied by the distance in the direction of that force. To carry up two sacks takes twice as much work; to carry them up six flights takes again twice as much. In the process of doing work on a system, we increase its energy.

Suppose, now, that we are dealing with forces that are *conservative*. Such forces are characterized by the fact that if we move a particle from some point A to another point B, the work we have to do does not depend upon the particular path chosen to get from A to B. The gravitational force is of that kind, and so are most other forces physicists deal with. If you walk to the top of a hill, the total work you have to do does not depend on which path you take. (The steeper path, of course, requires more work per mile covered and thus is harder, but if you want to exert the same total effort, you compensate for that by walking more slowly.) If you climb slowly, so that stopping requires negligible effort, the work you have put in has raised your *potential energy* by the same amount; the rise in potential energy depends only on the altitude gained, that is, on your new position, and not on how you got there. You can cash in your heightened potential energy by coasting to the bottom on a bicycle, at which point it has all been converted (assuming that there is no friction) into energy of motion, called *kinetic energy*.

Most of the forces we encounter in everyday life are not, in fact, conservative. However, as happens frequently in science, it is very much more fruitful to make a simplifying assumption (that forces are conservative), to build a theory on that basis, and subsequently to study the consequences when the assumption does not hold. Galileo's claim that bodies fall with a constant acceleration which is independent of their mass also does not hold exactly for most objects observed in everyday life; yet that assumption, which is correct under idealized conditions, led to enormous advances in physics.

If a system of particles contains only such conservative forces, the *total energy* of the system, that is, the sum of the potential and kinetic energies, can be expressed in terms of the positions and momenta of all the particles. The momentum of a particle is (in most practical cases) simply its velocity multiplied by its mass. For example, if we

DIFFERENTIAL EQUATIONS

Most of the laws of physics, including Newton's second law of motion, are expressed in the form of differential equations, which employ functions and their derivatives. A differential equation for some function $f(x)$ is an equation that contains not only $f(x)$ but also its derivative, and perhaps its second derivative and higher derivatives. (In fact, it may not contain $f(x)$ itself at all but only its derivatives.) If no derivatives higher than the first appear in it, it is called "of the first order"; if the second derivative appears, it is "of the second order," and so on.

A simple example of a first-order differential equation is given by $f'(x) = b$, where b is some specified number. The equation says that the slope of the plot of $y = f(x)$ is everywhere equal to b; its rate of increase is b. Therefore the graph (see Figure 3) must be a straight line with slope b, but its intersection with the y-axis is quite undetermined. In other words, the solution of the differential equation is $f(x) = bx + c$, where c is an arbitrary constant.

This result is an example of a general fact: the solution of a differential equation of the first order contains an arbitrary constant. One way of determining this constant is by supplying the differential equation with a *boundary condition*. (If the independent variable is the time, and the boundary condition refers to the initial time $t = 0$, it is also called the "initial condition.") This means that we specify the value of the solution $f(x)$ at some given point, say, $f(0) = 1$. In the above example, the solution is then $f(x) = bx + 1$. The solution of the differential equation with its boundary condition is generally uniquely determined. If we are given the speed $f_t = v(t)$ of a car along a given road as a function of the time, and we are also given the driving time, we still cannot tell where the car will stop unless we are also given its starting point. To specify a car's gas consumption per mile does not determine the amount of gas in our car's tank after a given number of miles unless we also know the amount of gas in the tank when we started. The solution of the system, consisting of the differential equation $f_t = v(t)$ and the initial condition that tells us the location of the car at the beginning, is then uniquely determined; without the initial condition, it is not.

If the differential equation is of the second order, two boundary or initial conditions are required to solve it uniquely. For example, the second-order equation $f''(t) = 0$ may be converted into a first-order equation by substituting $g(t) = f'(t)$, where g is an unknown function; the equation then becomes $g'(t) = 0$ for the function $g(t)$. The equation $g'(t) = 0$ for all t implies that the slope of g is always flat (its rate of increase is zero), and we may conclude that the graph of the function $g(t)$ is a horizontal straight line, $g(t) = c$, where c is an arbitrary constant number. It follows that $f'(t) = g(t) = c$, and as we saw earlier, the solution of this differential equation is $f(t) = ct + b$, where b is another arbitrary constant. Therefore, the solution of the original equation $f''(x) = 0$ contains two arbitrary constants, and these may be determined by the initial conditions $f(0) = k_1$ and $f'(0) = k_2$; in other words, f should have the initial value k_1 at the starting time $t = 0$, and it should have the initial slope k_2. The solution is therefore $f(t) = k_2 t + k_1$. If we are given the acceleration of a car as a function of the time, we can determine the point it reached during a given time interval provided we are also given its starting point and its initial speed. Again, this is an example of a general statement: if a motion satisfies a second-order differential equation and if we specify both the initial position and the initial velocity, then, although the equation by itself has infinitely many solutions, its solution is uniquely determined by the initial conditions.

A differential equation may also contain several dependent variables, and there may be several such equations that we must solve simultaneously. In that case, we speak of a *set* of differential equations, and boundary or initial conditions have to be supplied for all the dependent variables.

ignore the internal energy contained in its batteries and electronic apparatus and regard it as a "particle," the energy of a satellite in orbit around the earth can be expressed in terms of its momentum (that is, velocity times mass) and its position at a given time; the energy of the system consisting of the sun, the earth, and the moon (each idealized as a point particle and the rotation around their own axes ignored) can be completely specified by their positions and momenta at a given time. In the first example, the total number of variables is six (three for the position and three for the momentum), and in the second example it is eighteen (nine for the positions of the three objects and nine for their momenta). This energy, considered as a function of all the components of the positions q_1, q_2, ... and of the momenta p_1, p_2, ... of the particles, is called the Hamiltonian function, or simply the Hamiltonian, denoted by the letter H. To specify a given system of n particles, with all the forces acting on them and among them, is equivalent to specifying its Hamiltonian as a function of q_1, q_2, ... and p_1, p_2, ... ; we may simply say that H is a given function of $6n$ variables, the q's and the p's.

The Hamiltonian equations of motion are a set of $6n$ differential equations of the first order for the $6n$ dependent variables q_1, q_2, ... , and p_1, p_2, ... , all of which are functions of the time. These equations have the same content as, and are equivalent to, those of Newton (though they admit of greater generalization). They are simply reformulations that make Newton's equations easier to solve and that allow us to view the development of a system more intuitively. The equations of motion now contain only first derivatives, and in such a way that the time derivatives of the components of the positions and momenta are given functions of all the positions and momenta.

In other words, if, at the time t, we know the positions and momenta, then we also know their time derivatives. This implies that we can determine the momenta and positions a short time T later, because $f(t + T) - f(t) \simeq f_t\, T$, as we saw earlier, which amounts to approximating the curve for a short time by a straight line. We can therefore intuitively understand how such equations are solved, or "integrated," numerically by a computer: we start with the initial values of the dependent variables as initial conditions at the time $t = 0$ and increase the time by the small increment T; the above formula allows us to compute the dependent variables at the somewhat later time T, and we use these new values as initial conditions at the time $t + T$ to repeat the procedure. Many repetitions of these small steps eventually bring us to any desired later time and to the values of the q's and p's at that time.

The great intuitive advance of the Hamiltonian formulation of Newton's equations of motion is that in his version, once we know the p' and q's at one time, we also know their first time-derivatives at that time, and we can solve the equations.

Phase Space

Now that we have a description in terms of positions and momenta, it is intuitively very helpful to consider the motion of a system of particles in a space in which a point is identified by the positions and momenta of all the particles; this is called the *phase space*. A point in this phase space specifies not only where all the particles are at a given time but also what their momenta (or velocities) are. Laplace's "respective situation of the beings who compose it" has been reduced to specifying a point in phase space.

For example, if we are dealing with one particle moving along a line, the phase space is two-dimensional, with the particle's position plotted along one axis, and its momentum along the other, as in Figure 7 shown later. If the particle is free to move in three dimensions, we must specify the three components of $\vec{q}$ and the three components of $\vec{p}$. These add up to six numbers: the phase space is six dimensional. For n particles its dimension is $6n$; this simply means that $6n$ numbers are needed to specify a point. For the example, given above, of a pointlike satellite in orbit around the earth, the phase space is six-dimensional, and for the sun-earth-moon system it is eighteen-dimensional. In contrast with the trajectory of a particle in ordinary space, its path in phase space shows not only its position at various times but also its momentum.

So if we want to specify the present *state* (by which physicists mean the specification of the momenta and positions of all the particles that make up the system) of an n-particle system, we do it by specifying a point in its phase space. The motion of the system in the course of time is described simply as a curve or trajectory in its phase space. Such a curve completely describes the motion, even if the system is a gas consisting of 10^{23} molecules, so that its phase space has 6×10^{23} dimensions. Impossible as this may be to visualize literally, the idea is clear enough.

There is one limitation on the motion of a system point in its phase space, and that originates from the fact that energy is conserved. (For a discussion of why energy is conserved, see Chapter 10.) This is a result of the structure of the equations of motion for conservative forces; it is not something added to the equations. Remember that for a given system the Hamiltonian H is the energy and it is a known function of the p's and q's. To say that the energy is constant is to say that the p's and q's are constrained to move in phase space in such a way that, even though they all change in time, H remains constant. Geometrically, this means that the point in phase space must remain on a surface whose dimension is one unit lower than that of the phase space.

For example, if a point in three-dimensional space with the coordinates (x,y,z) is constrained so that always $x^2 + y^2 + z^2 = 1$, it must lie on the sphere of radius 1 about the coordinate origin, the zero-point. If, instead, $x^2 + y^2 + z^2 = 2$, then it must lie on a concentric sphere of radius $\sqrt{2}$. The reason for this is that in three dimensions (as you find by two successive applications of the Pythagorean theorem), $\sqrt{x^2 + y^2 + z^2}$ is the distance of the point (x,y,z) from the origin, and this distance is the same for all points on a sphere centered at the origin.

The equation $H = E$ specifies the surface in phase space on which the point representing the given system of energy E is constrained to move. If the same kind of system (with the same forces) has a different energy, its point lies on a different surface and remains on it. Two such surfaces can never intersect, since that would mean that the same system has two different energies. You may visualize the energy surfaces as resembling the layers of an onion, which might be infinitely large.

What kind of surfaces are these? Suppose the system does not have enough energy either to allow any of the particles to move infinitely far away or to let them have unlimited velocities. Then the system must perforce remain in a bounded region of phase space; it cannot escape. If such a system is one-dimensional, so that its phase space is two-dimensional, one can prove that, in fact, its trajectory must necessarily be a *closed curve* (see Figure 7). This implies that, since the system has to repeat its loop in phase space in an identical manner over and over again, its motion is necessarily *periodic,* like that of a pendulum.

How is the curve that describes the motion of a system in phase space determined? If the interparticle forces are known, so is the total energy at any instant as a function of the p's and q's; hence so are Hamilton's equations of motion. These first-order differential equations allow us, in principle, to find a system's trajectory in phase space once we know its starting point, that is, its initial conditions. Therefore, for a given Hamiltonian, that is, for a given system subject to given forces, only one trajectory can pass through a given point. No two trajectories of a given kind of system can ever cross one another, nor can a trajectory ever cross itself. Were it to do so, it would mean that this initial condition together with the differential equations would not determine the solution uniquely.

We are now in a position to understand, and perhaps even visualize in our mind's eye, the basis of Laplace's claim that was quoted at the beginning of this chapter. Even for the universe as a whole we may construct a phase space and imagine watching the trajectory of the

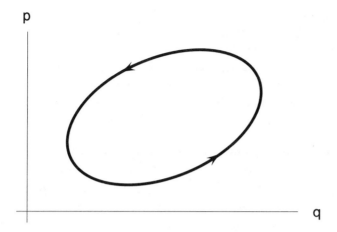

p

q

FIGURE 7. A closed trajectory in two-dimensional phase space.

world in the course of time, beginning at an arbitrary initial point. Its future course is forever determined, and if we are only intelligent enough, and we know the precise location of that point in the phase space of the universe, which means that we know its present state exactly, we can predict its motion as far in advance as we like.

So far, so good. We may now envisage implementing Laplace's project on a powerful supercomputer. At this point, however, we notice what at first appears to be no more than a minor difficulty. Since a computer is a digital device of high but not infinite precision, whatever data we feed in as "the present state of the universe" will differ from the acual state by a very small but unknown amount. It therefore behooves us to examine what the effect of such a small error on the accuracy of our prediction will be. In order to understand the consequences of small changes in initial conditions, we will have to look more closely at some specific physical systems, and we shall find that though Laplace was right (as far as classical mechanics is concerned), his vision was ephemeral.

Hamiltonian Flow

Let us return to our example of a system in one dimension (or with "one degree of freedom," as physicists say) with attractive forces; as we saw earlier it has a closed trajectory in phase space and must be periodic. Next, take a system with two degrees of freedom, for example, a ball that is free to move in a plane, with springs that pull it in two directions. Its motion in the x-direction is governed by one pair of springs, and its motion in the y-direction by the other pair; and these two motions are independent of one another (when the displacements are small). The phase-space in this case is four dimensional, but the resulting trajectory will lie on a two-dimensional surface because it is not only the total energy that is conserved but separately the energy of the motion in the x-direction and that of the motion in the y-direction. Furthermore, since we must again have periodicity of the x-motion and also of the y-motion, the surface is a *torus,* like the surface of a bagel (not necessarily of circular cross section; see Figure 8). Such a surface contains two kinds of closed curves: those around the hole and those perpendicular to the hole. If the two pairs of springs are equal, then the x and y periods are equal—say T—and the system returns to its original point in phase-space after one period T. The entire motion thus has the period T. If the x-period T_x and the y-period T_y stand in the ratio $T_x/T_y = 5/3$, since $3T_x = 5T_y$, three repetitions of the x-motion or five repetitions of the y-motion return the system to its original point. It is still periodic, with the period $T = 3T_x = 5T_y$. Whenever the two periods are commensurable, that is, their ratio equals the ratio of two whole numbers, the resulting motion is periodic and the trajectory on the torus is a closed curve (see Figure 9 for an example). If the two periods are not commensurable, the trajectory is not closed; in fact it will cover the entire surface of the bagel and will eventually pass through any chosen patch on the surface, no matter how small. The motion is now not periodic but *quasi-periodic.* Tracing the motion of the particle in the (x,y)-plane rather than in phase space, we observe, in the periodic case, a complicated but beautiful curve called a *Lissaj-*

FIGURE 8. A two-dimensional torus.

CHAOS AND THE GHOST OF LAPLACE

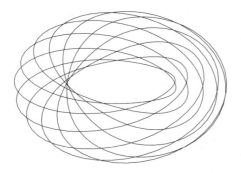

FIGURE 9. A closed trajectory in four-dimensional phase space on a two-dimensional torus.

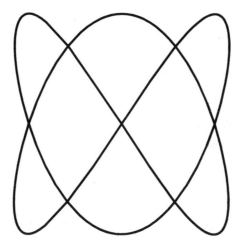

FIGURE 10. A Lissajous figure.

ous figure (see Figure 10), after the French physicist Jules Antoine Lissajous, who first produced such curves by optical means from the vibrations of tuning forks.

Integrable Systems

These notions carry over to more complex systems with more degrees of freedom. Whenever a suitable set of pairs of variables can be found for a system of n degrees of freedom (whose phase space is $2n$-dimensional), such that each pair is a periodic function of the time, the motion is periodic if all of the individual periods are commensurable, and quasi-periodic otherwise. The phase-space trajectory is then confined to an n-dimensional torus, and in the quasi-periodic case it will densely cover the toroidal surface. Even so, since the dimensionality of the whole energy surface is $2n - 1$, which for n greater than 1 is larger than n, the trajectory is much more constrained than it would otherwise be. Such systems are called *integrable*. The property of integrability usually results either from some specific symmetry property of the system or from some other special property it possesses.

An example of what constitutes a symmetry property would be when the forces in the system are invariant under a rotation of the whole. By other special properties I mean, for instance, that some parts of the system may be independent of others, so that the energies of the separate parts are individually conserved. For instance, take two separate pendula that do not interact with one another. The entire system consisting of the two pendula will be integrable; since each of the two is periodic, the whole system is either periodic or quasi-periodic, depending upon the commensurability of the two periods. The phase space of this system of two degrees of freedom is four-dimensional and its energy "surface" is three-dimensional. But since each of the two energies is separately conserved, the phase-space trajectory lies on the surface of a two-dimensional torus.

Consider now what happens if we take such a system of two pendula and we add a mechanism by which the two become coupled in some simple way, that is, the motion of one begins to influence in some specified manner the motion of the other. For instance, the bob of one pendulum may be connected to that of the other by a spring. In general, one expects that the integrability of the system will be lost by such coupling, because the special property of consisting of two independent subsystems, each of whose energies is separately conserved, is lost. This means that whereas without the coupling the trajectory in phase space is confined to the surface of a (two-dimensional) torus, with the coupling present the trajectory will be able to wander all over the (three-dimensional) energy surface. At least, this is what used to be expected on the basis of the knowledge of classical mechanics as it was developed from Newton's equations in the eighteenth and nineteenth century by the great mathematicians Leonard Euler, Adrien Marie Legendre, Carl Gustav Jacobi, Henri Poincaré, and others (as well as Lagrange, Laplace, Liouville, and Hamilton, whom we have already mentioned).

CHAOS AND THE GHOST OF LAPLACE

Transition to Nonintegrable Systems

The existence of high-speed computers has made it possible to investigate extensively the kind of mechanical configuration I have just described, two pendula with a simple kind of coupling between them, as an example of a system that might be thought of as "almost integrable." It was first introduced as an instructive model for the study of more complicated systems by two astronomers, Michel Hénon and Carl Heiles, in 1964. Since the equations of motion cannot be solved explicitly even for this simple system, their solution had to be found numerically. The result was quite surprising and is still mathematically not fully explained. In order to comprehend the puzzling result, we must look at a method of determining in a simple way whether a system such as this is integrable.

If we intersect the three-dimensional energy surface with a plane that slices through the bagel to which its phase-space trajectory is confined if it is integrable, it is easy to see whether a system of two degrees of freedom is integrable. This plane is called a "surface of section." The trajectory of the system will pierce the surface of section every time it comes around, and we mark the point of its intersection. If the trajectory lies on a torus, the marked points of intersection will lie on a recognizable closed curve, the intersection of the surface of the bagel with the surface of section. These points do not form a curve in any orderly way or by a regular progression in time, but after many intersections have taken place, the curve formed by the marks is easily recognizable as such. On the other hand, if the system is not integrable, the intersection points wander chaotically all over the surface of section.

What Hénon and Heiles found by computer calculations is that for low energies the system they studied appears to be integrable, but as the energy is increased (or the coupling between the oscillators is strengthened) the motion becomes chaotic. In other words, for low energies or weak coupling, the intersection marks on the surface of section, after many transits, form a number of clearly visible closed curves, which show the outlines of the intersections of tori with the surface of section. But for higher energies or stronger coupling, an increasing number of the marks wanders all over the surface of section in a random manner, and eventually all of the marks become chaotic. Figures 11, 12, and 13 are computer printouts of surfaces of section for the system studied by Hénon and Heiles, at higher and higher energies.

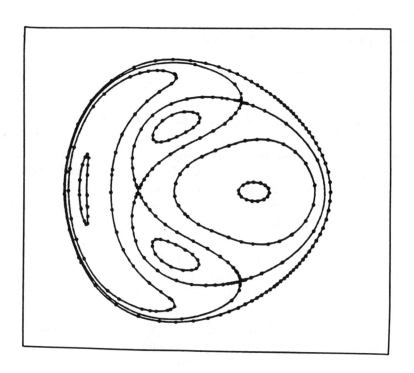

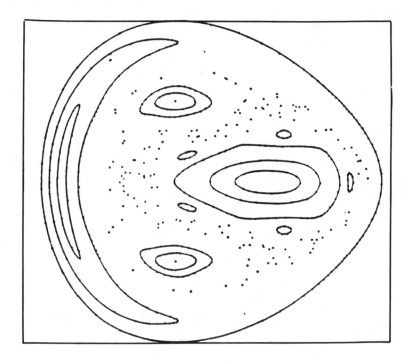

CHAOS AND THE GHOST OF LAPLACE

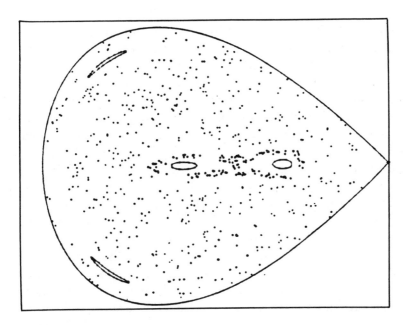

FIGURES 11, 12, 13. A surface of section for the Hénon-Heiles system at low energy, at a higher energy, and at a still higher energy.

There is still no full understanding of this behavior; however, there is a remarkable theorem that is thought to give a general, qualitative explanation. It was discovered forty years ago by the mathematician Andrei Kolmogorov, and its proof was refined by Vladimir Arnold and Jürgen Moser. This so-called *KAM theorem* states that, contrary to what used to be believed by great experts such as Poincaré, if a system with a sufficient number of symmetries to make it integrable is perturbed so that these symmetries are destroyed, the system does not necessarily become nonintegrable. For a certain range of perturbations and energies the integrability may be preserved, even though its putative cause is gone. The theorem, however, does not tell you how to determine the kind and strength of the perturbations for which the system remains integrable; it is "nonconstructive," that is, it asserts only their existence. The Hénon-Heiles system is believed to be an example of this KAM behavior.

Chaos

The way the situation is to be generally understood is that there are, in nature, integrable systems, whose motion exhibits a certain specific kind of orderliness, and others, whose trajectories in phase space are chaotic.* The overwhelming majority of systems are of the second kind, even though the KAM theorem tells us that the former kind is not quite as rare as physicists used to believe. If a system is not integrable but "nearly" so, there may still be certain exceptional initial conditions that lead to periodic or quasi-periodic motion, but almost all initial conditions will not. Eventually, when a system is sufficiently "far away" from being fully integrable, and no initial conditions lead to periodic or quasi-periodic motion, its "Hamiltonian flow," that is, the flow of its trajectories in phase space, is generally what is called *ergodic*. This means that a given trajectory will not only wander all over the energy surface, but it will come arbitrarily close to every point on it. If we surround any given point by a sphere whose radius is as small as you like, and wait long enough, the trajectory is sure to come through it. This means that any two systems of the same kind with the same energy will, at some time in their evolution, have states that are almost identical. Indeed, if you divide up the energy surface (whose total volume is finite for systems of attractive forces and negative energy) into small patches of equal volume, the trajectory will eventually spend equal lengths of time in each patch. In a certain sense each state that is accessible to a system with a given energy is equally probable.

Suppose that two identical physical systems of many degrees of freedom start out with initial conditions that are very nearly the same. What will generally happen is that while their trajectories remain close to each other for a period of time, after a long time their points in phase space will become widely separated. In fact, their states will, for all practical purposes, be indistinguishable from systems of entirely different initial conditions. They appear to have lost all memory of their initial conditions, even though we could, of course, trace them back to their initial points by following their trajectories backwards.

The behavior of systems with different initial conditions may be visualized by taking a small chunk of phase space and letting each point in it evolve in time according to the same Hamiltonian. Its shape will become distorted, as in Figure 14. However, according to a theorem proved by the French mathematician Joseph Liouville, the

*It is no accident that we have not defined exactly what is meant by "chaotic"; such motion is hard to describe precisely but relatively easy to recognize when you see it.

CHAOS AND THE GHOST OF LAPLACE

FIGURE 14. A volume in phase space gets
distorted in the course of time.

volume of the chunk remains unchanged as though it were an incompressible fluid. In spite of this preservation of its volume, however, if the system is nonintegrable and chaotic, as almost all real systems are, the chunk will eventually spread all over its energetically allowed region in phase space. If it was initially nice and compact, say a neat round ball, after a while it becomes an octopus whose tentacles wind themselves all over the place, even though it still has the same volume (see Figure 15). As a result, two points originally close together in the ball may wind up at the ends of opposite legs of the octopus and thus very far apart. No matter how close two identical systems were initially, they will, after a long time, differ enormously from one another.

Recurrence and Time Reversal

The chaotic behavior of most systems and the wide separation of points in phase space that were initially close together become even more surprising if we add what is known as Poincaré's *recurrence theorem*. (Since Laplace lived a century before Poincaré, he did not know this theorem, which adds a certain poignancy to his vision.) For a system with a finite energy surface in phase space, Liouville's theorem implies that any given little volume about an initial point, as it sweeps around, must eventually exhaust the whole allowed volume of phase space; it cannot keep wandering around without overlapping some region it already covered. This must first happen near the initial position. Therefore, in the course of time the system *must return arbitrarily closely to its initial state*. But this does not imply that all the other points in the initially designated small ball near the starting point must also return at the same time, causing the long, thin ribbon into which the ball had developed to retract again into a ball. While one system has returned to its home plate, others that

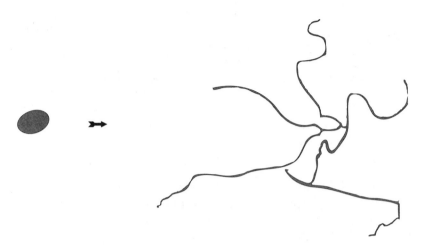

FIGURE 15. The long-time development of a bit of phase space.

started in the same vicinity may at the same time be out in left field, though each must, at one time or another, return home. The recurrence times for two points may be vastly different, even though they started very near to one another. They are said to be "extremely sensitive to the initial conditions." These recurrence times for natural systems of many degrees of freedom are enormous; for the molecules in a beaker of gas, the recurrence time is very much longer than the age of the universe. For the world as a whole, we have something resembling a literal version of Nietzsche's "eternal return," but with a small variation. Note that since the world will not return *exactly* to its initial state, and since its trajectory is sensitive to its initial conditions, its history the second time around will not resemble that of the first time.

A second apparent puzzle arises from the fact that if the sense of time is reversed, so that the motion runs backward, the Hamiltonian (or Newtonian) equations of motion change in the same way as they do if all the momenta are flipped around, so that the motion runs forward but all velocities in it are reversed. In other words, if you make a movie of the development of a system and you run the movie backwards, what you see is a perfectly legitimate motion; there is no way for you to be able to tell that the movie is shown backwards. (There is no friction in the systems we are talking about; nothing runs down and no energy is lost.)

Imagine taking a movie of a flow in which the "fluid" in a small ball is colored red. You see the fluid from the red ball distributed

CHAOS AND THE GHOST OF LAPLACE

more and more widely as time goes on, and eventually it becomes a tangle of a thin red ribbon that seems to fill most of the available phase space; you stop and run the movie backwards, which shows a development of the system that is just as lawful as the forward motion. But in this showing of the film, what you see is the thin red tangled ribbon untangling itself and finally retracting into a neat little ball! In the forward motion the development moved from what appeared to be a well controlled and orderly state to a chaotic mess; in the backward motion the chaotic tangle develops into neat order. Both follow the same equations of motion.

We have to conclude that what appeared to be total chaos may not have been quite as chaotic in all respects as it seemed, just as the mess on top of my desk that appears to my wife to be an impenetrable confusion of papers is really not as confused as it seems; I can quickly find any needed buried letter because I can remember where I put it. (This metaphor is a bit stretched; the apparent mess on my desk is, in fact, a real mess and I often cannot find important letters in it.) This does not mean, however, that every such tangle can be equally untangled by letting the system run long enough in one direction of time or the other. If one of the ribbons of the "ordered mess" is shifted ever so slightly, the tangle will never get back into the neat little ball from which it started. Instead, at least a part of it will always remain a tangled ribbon. Again we have "sensitivity to initial conditions," on the way back as well as on the forward journey.

Laplace's Dream Fades

So what has become of Laplace's perfectly predictable clockwork universe? As far as classical mechanics is concerned, he was, in principle, of course entirely correct. However, this principle could not be implemented even on the largest imaginable computers. There can never be perfect correspondence between even the most precise knowledge of initial conditions and the use a computer can make of that knowledge, since computer calculations must, at every stage, utilize numerical approximations. Every number used by a computer at every stage of its calculations has a finite number of decimal places and is thus almost always an approximation. The tiniest initial deviation or calculational round-off will eventually lead to large differences between the real course of a system and the computed result. Such calculations of the development of a system, if meant to extend into an indefinite future, are therefore of limited utility, and predictions based on them are bound eventually to go wrong. What is more, "eventually" often comes about quite soon. Attempts at long-range weather forecasting are a good example. John von Neumann, one of the pioneers of the modern electronic computer, expected computers to enable meteorologists to make reliable long-range forecasts. This expectation has not been fulfilled; and because the prevalence of chaos is today more fully appreciated than it was in von Neumann's day, very few people now expect it ever to be.

During the last ten years, the study of chaos in deterministic systems has become very fashionable, and some of its enthusiasts pretend that this area of physics is quite new. In fact, the existence, indeed the prevalence, of such systems in nature has been known for a hundred years, at least since the work of Poincaré. In the past, however, physicists generally preferred to study the orderly, albeit exceptional, systems, because they could be more readily understood. That now many scientists prefer to think instead about chaotic systems may possibly be a psychological reflection of the social chaos of our time, and an example of the influence that cultural factors exert on science, which a number of sociologists of science have propounded and sometimes greatly exaggerated. It is also undoubtedly because high-speed computers have made numerical investigations of such systems technically feasible; in Poincaré's day such computations would have taken far too much time to be practical.

The principle of Laplace's dream of deterministic order remains like the smile of the cheshire cat, lingering in the air while the cat has vanished. There is, however, an area of physics in which this principle puzzlingly coexists with quite different observed facts, leading to something entirely new: systems that consist of very large numbers of particles in chaotic motion, like gases and fluids, behave *irreversibly*. Though a videotape of the motion of each molecule separately could be reversed without anyone being the wiser, a movie of an ink drop being mixed into a glass of water, if run backwards, would be immediately recognized as an "impossible" course of events. Why this is so we shall examine in the next chapter.

3

TIME'S ARROW

The objects whose motions mechanics is designed to understand and describe are point particles and either rigid or deformable bodies; in the last chapter we discussed the concepts and mathematical tools developed during the last 300 years for this purpose. In order to understand the motion of such objects, we have to know their mass, shape, elasticity, and the forces they are subject to. (For simplicity we confined ourselves to point particles, so that shape and elasticity were not relevant, and we did not go into the more complicated problems that arise for extended bodies.) But there are other important physical properties of the objects around us; the one that we are probably most aware of is temperature.

We all know the difference between a hot and a cold substance. Furthermore, we know very well that a hot cup of coffee left standing around grows cold; its temperature eventually becomes the same as that of the ambient air. If you wrap your cold hands around it while it is still hot, they are warmed: heat flows from the hot cup to your cold hands. Have you ever found that when you wrapped your hands around a hot cup of coffee your hands became colder and the coffee hotter? No one has ever observed such a thing. If someone were to show you a video tape that depicted a dark iron bar dipped in a bucket of boiling water, then emerging red hot with the water full of ice, you would immediately understand that the tape was running backwards. Such a course of events has never been seen. In startling contrast to the mechanical motions described in the last chapter, many phenomena involving the flow of heat are *irreversible*. It is the science of heat, its interaction with mechanical work, and the phenomenon of irreversibility that we want to discuss in this chapter.

Temperature

Already in the seventeenth century Isaac Newton announced a general law known as Newton's law of cooling: the rate at which a hot object cools is proportional to the difference between its temperature and the temperature of the air around it. When the coffee is very hot it cools rapidly, but as it gets cooler, its rate of further cooling is much slower, as you can easily verify. But what do we mean by *temperature*? We clearly need a more objective measure than the sensation in our hands, because we all have had the experience that a cold metal surface feels much colder than a wooden surface of the same temperature.

The first step in the definition of temperature is to recognize that whatever it is we want to mean by it, it should be such that when any two objects have been in close contact for a long time, their temperature is the same. Two such objects are said to be in *thermal equilibrium*. This is an important first step because we can then determine the temperature of any piece of material by bringing it into thermal equilibrium with a given standard measuring device that is somehow calibrated to show its temperature: a *thermometer*.

While the idea of a thermometer goes back as far as about 100 BC to Philon of Byzantium, the first modern thermometer was invented in 1593 by Galileo. The Dutch inventor Cornelius Drebbel thought of it independently at about the same time in order to regulate the temperature of furnaces and ovens. The principle on which it was based is the expansion of a liquid caused by a rising temperature; a fixed amount of the liquid (at first alcohol was used, but today we use the stabler mercury) would rise in a calibrated glass tube. The scale and the zero point remained still to be fixed. In 1664 Robert Hooke proposed the melting of ice as the standard zero point because he had noticed that a mixture of ice and water always gave the same reading of a thermometer; thirty years later the Italian mathematician Carlo Renaldini chose the boiling point of water as the other reference point and divided the interval between the freezing and boiling of water into twelve equal parts. (Meanwhile the astronomer Joachim Delancé, being French, had proposed as an alternative zero point the melting of butter, which however was rejected as too unreliable.) The final choice, which we use today, was made by the Swedish astronomer Anders Celsius, who divided Renaldini's interval into a hundred equal parts. This scale—also called the *centigrade* scale—is named after him.* (Actually, his orginal proposal was to designate the boiling point of water as 0° and the freezing point as 100°; this was later turned around.)

The zero point of the Celsius scale, however, still has a certain arbitrariness about it; the freezing point of water, practical though it is to locate, has no general scientific significance. However, the combined experimental results of Boyle in the seventeenth century and of Charles, Dalton, and Gay-Lussac in the eighteenth led to the conclusion that while for an "ideal gas" (any gas at low enough pressure is "ideal") at a constant temperature the product of its pressure p and volume V has a fixed value, this value varies linearly with the temperature t. More specifically, if we take a fixed number (called Avogadro's number) of molecules of any ideal gas, then $pV = R(t + a)$, where a is a certain constant, and R is called the *universal gas constant*. If the temperature t is measured in degrees Celsius, the constant a is found to have the value 273.16°. Therefore, introducing a new temperature scale $T = t + a$, we find that the "ideal-gas law" simply reads $pV = RT$. This scale is called the "ideal-gas temperature," the *absolute temperature,* or the *Kelvin* scale after Lord Kelvin, who made many important contributions to physics. Each degree on the Kelvin scale, the measure of temperature that is now used universally in the science of heat, is of the same size as a centigrade, but its zero point is at −273.16° Celsius.

Heat

Our discussion at the beginning of this chapter opened with the early observation that hot bodies raise the temperature of cold ones in

---

*The Fahrenheit scale was proposed by Daniel Gabriel Fahrenheit, who used a mixture of ice, water, and sea salt for his zero point and designated the human blood temperature as 96° (now more accurately known to be 98.6°); this scale is not used for scientific purposes. Since water freezes at 32°F and boils at 212°F, you convert the temperature in the Fahrenheit scale into centigrades by subtracting 32, dividing the result by 9, and multiplying it by 5.

contact with them. The nature of the heat that is obviously being transferred, however, remained a mystery. For the ancient Greeks, fire was one of the four elements of nature, and even in the first half of the nineteenth century the word fire was used as a synonym for heat. A more specific idea arose in 1760 when the Scottish physicist and chemist Joseph Black introduced the notion that heat was an indestructible fluid filling the interstices of solid bodies, liquids, and gaseous materials and having the intrinsic property of flowing from a higher temperature to a lower one, just as water flows from a higher level to a lower; this became known as the *caloric theory* (the word comes from the Latin "calor" for heat) of heat, particularly popular in England for the next three quarters of a century.

Forty years later, however, a competing theory was introduced which became known as the *vis viva theory* or *kinetic theory* of heat. (*Vis viva* was a name used at the time for the energy of motion of an object, what we now call kinetic energy.) Benjamin Thompson, an American born before the revolution, who moved to England and eventually became Count Rumford, had noticed the large amount of heat produced in the boring of cannon barrels and concluded that heat must be a form of vibratory motion of the constituents of matter. The English chemist Humphrey Davy came to the same conclusion about 1803 by rubbing two pieces of ice together, thereby melting them. Exactly what it was that was vibrating was not always agreed upon. Although already in 1738 Daniel Bernoulli* had proposed that the pressure exerted by a gas was caused by the momenta of its fast-moving molecules bouncing against the walls of its container, not everybody agreed with this model. Others pictured a gas more like a solid, and the heat-motion as small vibrations. Nonetheless, the caloric theory was quite distinct from any version of the kinetic theory, which we now accept as correct in the form given by Bernoulli. During the first half of the nineteenth century, however, the two competing visions of the nature of heat existed side by side, especially in France.

Let us now turn to the real action of heat. In 1765 the Scottish engineer James Watt had invented the condenser for the steam engine, which made that engine, both literally and figuratively, the driving

---

*The Bernoulli family was the Bach family of science. Daniel was a member of the second of three generations of men, eight of whom—his father, his uncle, two brothers, a cousin, and two nephews—made significant contributions to mathematics and physics.

force behind the Industrial Revolution in Europe and America, with far-reaching social and technological consequences. Here was an enormously practical and useful method of converting heat into mechanical work. The men who developed the science of thermodynamics, which deals with that conversion, sought not so much to improve it as to understand it; the steam engine driving most of the machines in the ever-expanding industries was their basic starting point. Both Sadi Carnot, a French engineer whose father had been Minister of War under Napoleon I, and Rudolf Clausius, a German physicist, two of the principal contributors to thermodynamics, opened their main works on the subject by referring to the steam engine as the agent for converting heat into motive power. Clearly, once it had been established that mechanical work could be converted into heat and *vice versa*, the caloric theory, which postulates that heat by itself is conserved, could not long survive. Nor did it, although much of the early development of thermodynamics occurred at a time when the caloric and the kinetic theory coexisted. Thermodynamics, however, is a phenomenological theory and does not, as such, depend on an underlying model of the nature of heat.

The First Law of Thermodynamics

Heat can be converted into work, and work into heat. But work is measured in mechanical units, in terms of a force exerted over a certain distance, while the natural way of measuring heat is to determine the number of degrees by which it increases the temperature of a standard piece of material. This raises the question of the *mechanical equivalent of heat*: how much work is equivalent to how much heat? It was first measured roughly by Thompson in 1798, and with some precision in the years 1845 to 1847 by James Prescott Joule. Joule's famous experiment consisted of a set of paddles rotating in a bucket of water, which gradually raised the water temperature; measuring both the mechanical energy expended in the rotation and the rise in temperature, he established the mechanical equivalent of heat in a quantitative way, allowing us to conclude that the sum of the two energies, the mechanical energy and the heat energy, is conserved. Similarly for electrical energy: by driving his paddles with an electric motor, Joule found the equivalence between heat and electrical energy. Conservation of the total energy—that is, of the sum of the mechanical, electrical, and heat energies—is called the *first law of thermodynamics*. While Joule is credited in part with its discovery, the history of this law is somewhat complicated.

Before anyone else, the German physiologist Robert Mayer, who was serving as a ship's doctor on an extended voyage in 1842, was led to the equivalence of mechanical and heat energy by observing that the blood of European sailors was much redder when they were in the tropics than at home. From this he concluded that less work for the extraction of oxygen from the blood was required to keep the body at its normal constant temperature in the tropics than in cooler climates. By a chain of somewhat convoluted reasoning, he arrived at the conservation law, though he had to defend his priority against Joule for some time. A similar line of thinking was used by Hermann von Helmholtz, a man of many scientific accomplishments, who argued, on the basis of his own experiments, that body heat and muscle force originated from the oxidation of food. Thus the "vital force" postulated by others at the time as responsible for sustaining life was not needed and did not exist. Both Helmholtz and Clausius, who argued systematically that Joule's experiments proved heat alone could not be conserved but that the total energy was, are sometimes credited with the law of conservation of energy.

Many attempts were made in the nineteenth century and earlier, as well as occasionally even today, to construct an engine that would run forever, producing useful work without any source of energy. Such an engine, the industrial equivalent of the medieval philosopher's stone that was reputed to convert lead into gold, came to be called a *perpetuum mobile,* a perpetual motion machine. The first law of thermodynamics, stated negatively, denies the possible existence of such a machine, no matter how clever the inventor.

From our present perspective, in which we regard matter as made up entirely of particles and see heat as consisting of both random motions of these particles and electromagnetic radiation, the first law of thermodynamics is wholly subsumed under the energy-conservation laws of mechanics and electromagnetism (to be taken up in the next chapter), the deeper origins of which we shall discuss in Chapter 10. Although the first law played a very important role in the history of physics, it appears to modern scientists as somewhat redundant.

The Carnot Cycle

While we now know that heat and work can, in principle, be exchanged, how can we understand the manner in which the flow of heat is converted into mechanical work in an engine? Sadi Carnot's important contribution to the science of heat consisted in the construction of an idealized cyclical method for extracting work with the greatest possible efficiency from the heat-flow from one substance at a higher temperature to another at a lower one. This *Carnot cycle* is entirely independent of any underlying model for the nature of heat (he believed in the caloric theory) and of the detailed construction of the engine; it is a reversible process, which means it can run in either direction. Our interest in the *Carnot engine* does not stem from a fascination with engines as practical tools but from its use for schematic arguments in thermodynamic reasoning. Of course, that kind of argumentation betrays the historical origins of thermodynamics in the steam engine, but no satisfactory substitute, short of a complete mathematical axiomatization, has been found.

Imagine that a cylindrical container closed with a piston—a container whose volume can be changed by pushing the piston in or out—is filled with a fixed amount of gas at a certain given pressure higher than that of the outside air. (This is not really necessary, but makes it intuitively easier to understand.) Figure 16 shows a graph in which the pressure of the gas is plotted against its volume. At the beginning, the container is in contact with a "heat reservoir" (think of it as being immersed in a bathtub; the word *reservoir* is meant to imply that it is so large that heat can be extracted from it or be allowed to flow into it without any noticeable change in its temperature) at the temperature T_1 and is allowed slowly to expand, pushing the piston farther out and doing work (by raising a weight, for example), while its contact with the reservoir maintains it at the same temperature. Its pressure-volume point thus travels along the curve from A to B in the diagram. At the next stage, the container is taken away from the reservoir and put in an insulating jacket that prevents any heat from entering or escaping, the piston is allowed to slide even farther out, and the gas cools down to the temperature T_2; at the same time, it loses pressure and the point representing it travels from B to C. At the third stage the jacket is removed, the cylinder is immersed in a heat reservoir at the temperature T_2, the piston is pushed inward, and the gas moves along the curve from C to D. Finally, the cylinder is again removed from the reservoir and insulated, the piston is pushed inward to its starting position, with the pressure rising in the gas to its original value. At this point the gas must have the same temperature that it had at the beginning because its pressure and volume together determine its temperature either by

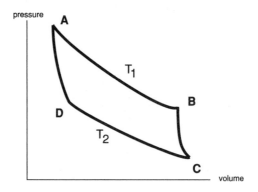

FIGURE 16. The pressure-volume diagram of a Carnot cycle.

the ideal-gas law if applicable, or by another "equation of state"; the gas is back at point A in the diagram. So the Carnot engine is now once more in its original state, but a certain amount of work W has been done by the piston. (The quantity W, it turns out, equals the area of the figure bounded by the four curves with vertices ABCD.)

The First law of thermodynamics requires that this work be accounted for by the net amount of heat absorbed by the gas during the periods when it was in contact with the reservoirs at the temperatures T_1 and T_2; during the other two parts of the cycle, the cylinder was insulated and no heat could enter or leave it. Let us call Q_1 the amount of heat that flowed into the cylinder when it moved from A to B, and $-Q_2$ the amount of heat that flowed into it (that is, Q_2 flowed out of it) from C to D. Conservation of energy implies that $W = Q_1 - Q_2$. The cycle may be repeated as many times as we like, producing mechanical energy by transporting heat from a source at a higher temperature to a sink at a lower one.

We may also imagine the cycle going in the reverse direction, with work being done on the piston and heat flowing from the reservoir at lower temperature to the one at higher temperature; this is a *Carnot refrigerator* if the purpose is to extract heat from the cool pool, or a *heat pump* if it is to pump heat into the hot one. In either case, it uses mechanical energy to move heat from a body at a lower temperature to one at a higher temperature.

How would we evaluate the efficiency of the Carnot engine? If all the heat extracted from the hot reservoir were converted into work, we would call it 100% efficient; more generally, the efficiency is the ratio of the work done to the heat extracted. But since we have seen that by conservation of energy $W = Q_1 - Q_2$, the efficiency differs from the ideal 1 (that is, 100%) by the ratio of the waste heat into the cool reservoir to the heat extracted from the hot one. The engine could be 100% efficient only if there were no waste heat-flow into the second reservoir. This, however, would violate Kelvin's version of the second law of thermodynamics, to which we now turn.

The Second Law of Thermodynamics

At the beginning of this chapter we noted that in our common experience we always find that when no work is done on a system, heat flows from a hot body to a colder one, and never the other way around. This heat flow can be used to extract work by a Carnot engine, and it can be reversed by the expenditure of mechanical energy, as we have also seen. The question arises, can an engine be constructed that performs mechanical work by extracting heat from a reservoir while allowing part of it to flow back into the same reservoir? Since the amount of heat energy in the earth's oceans and landmass is enormous, such a machine would be practically equivalent to a *perpetuum mobile*, even though it would not violate the first law of thermodynamics; it is called a *perpetuum mobile of the second kind*. The *second law of thermodynamics* says that such a machine is impossible. In the formulation of Kelvin the postulate states: *there exists no process whose only effect is to convert heat from a single thermal reservoir to work*. It implies that the process of producing heat by friction is irreversible. Coasting on your bicycle on a flat road, you will eventually come to a stop by friction producing heat in the tires, but without an expenditure of work; no process exists without energy expenditure that would accelerate your bike by cooling down those tires.

Clausius stated the same postulate in another way: *there exists no process whose only effect is to transfer heat from a cold to a hot thermal reservoir*. In this formulation the law says in effect that there can be no refrigerator that works without some other effect on the immediate environment (such as extracting energy from it). It implies that the process of heat conduction is irreversible; whereas no external energy is required for your warm house to cool down to the outside temperature in the winter, the furnace must use energy to heat it up again. The statements by Kelvin and Clausius sound different but are, in fact, equivalent, as we can see in the following way.

Suppose the Clausius postulate were false and we could, without any other effects, move an amount of heat from a reservoir at the lower temperature T_2 to one at the higher temperature T_1. We could then utilize a Carnot engine to extract that same amount of heat from the hot reservoir, converting part of it into work and dumping part of it back into the cold reservoir. Since the hot reservoir has had the same amount of heat that flowed into it extracted, it is unchanged, and we have converted heat from the cold reservoir alone into work, which violates the Kelvin postulate. Conversely, suppose that Kelvin's law were false. We could then obtain mechanical energy by extracting heat from a single reservoir at some temperature T_1, and subsequently converting that energy, say by friction, completely into heat, thereby raising the temperature of some other body whose initial temperature T_2 was higher than T_1. We would then have transferred heat from the colder reservoir to the warmer object with no other changes, which would violate the Clausius postulate. Therefore, these two ways of stating the second law of thermodynamics are equivalent.

At this point, one might ask what these postulates by Kelvin and Clausius, stated in the negative form that something cannot happen, have to do with versions of the second law found in popular discussions, such as the "increase of entropy" and even the "heat-death of the universe." To see the connection, we first have to define what entropy is.

Entropy

On the basis of his own early version of the second law of thermodynamics, Sadi Carnot proved that there can be no engine of any kind, reversible or irreversible, operating between the temperatures T_1 and T_2 (that is, letting heat flow from the temperature T_1 to the temperature T_2) to extract mechanical energy, that would be more efficient than his idealized one. From this, we may conclude that the maximal efficiency, that of the Carnot engine, can depend only on the two temperatures T_1 and T_2, and therefore the ratio Q_2/Q_1 must also depend only on these two temperatures. But then that ratio can depend only on the ratio of one function $f(T_2)$ to the same function $f(T_1)$, as you can see by replacing the single Carnot engine by two, the first operating from T_1 to T_0, and the other from T_0 to T_2, while keeping T_0 fixed. For the ideal-gas temperature scale, it turns out that this function $f(T)$ is simply T itself; this may, in fact, be regarded as the definition of the absolute or Kelvin temperature scale. Thus we have $Q_1/Q_2 = T_1/T_2$, which we may also write by multiplying the equation by Q_2, dividing by T_1, and rearranging, as

$$\frac{Q_1}{T_1} + \frac{-Q_2}{T_2} = 0.$$

(Remember that $-Q_2$ is the heat absorbed by the gas, which means Q_2 is the heat expelled by it, during the third part of the cycle.)

Now the ratio of the increase of heat of a body (counted negative if it is a decrease) to its temperature is called the increase of its *entropy,* an enormously fruitful concept invented by Clausius in 1865. (He chose the name "entropy" from the Greek word for transformation.) Therefore, the equation we have found says that for a (reversible) Carnot cycle, the total change of entropy is zero. Other kinds of reversible cycles do not have the simple property that all the heat is absorbed or expelled at a constant temperature. However, every such cycle can be replaced by a series of Carnot cycles, each of which extracts and expels only a small amount of heat while their temperatures are constant. Since for each of these Carnot cycles the change of entropy is zero, we may conclude that for any reversible cycle, the sum of all the small changes in entropy that take place add up to zero; in other words, the total change in entropy vanishes.

But if the net change of entropy of a body is zero as it goes through any reversible cycle that returns it to its starting point, then it follows that its change of entropy as it turns reversibly from one state to another does not depend on the manner in which it does so, that is, it does not depend on the path in the pV-plot that led from the initial point to the final point. This is because any two reversible paths differ from one another by the addition of a closed reversible cycle. (Just as wrapping a line from one side around a tree trunk to the other side along the left may be thought of as doing it along the right and then going back around the entire tree: the difference between the two ways is the closed path around the tree.) Therefore, since the entropy of a body does not depend on how the body arrived at its state, that entropy may be regarded simply as a property of the state, of its pressure and volume, for example, the only remaining ambiguity being the choice of some inital reference state that is fixed once and for all. It is, in this respect, quite analogous to the energy.

Suppose, now, that a small amount of heat Q is transferred from a heat bath at the temperature T_1 to a body at the temperature T_2; then the body's entropy increases by Q/T_2, while that of the reservoir decreases by Q/T_1. If body and reservoir are isolated from the rest of the world so that there is no source of mechanical power, the process is reversible if $T_1 = T_2$ and, by the Clausius postulate, irreversible if T_1 is greater than T_2. In the first case, the entropy of the body increases by exactly as much as the entropy of the bath decreases, and the total entropy remains unchanged; in the second case, the entropy of the reservoir decreases by less than the body's increase in entropy. Therefore, for the reversible heat transfer, the total entropy is constant, and for the irreversible heat transfer, it increases. For large heat transfers at varying temperatures, we divide up the total into small bits and draw the same conclusion.

On the other hand, suppose that an equilibrium system (that is, a system of uniform pressure and temperature) undergoes an irreversible process without heat transfer. We may then substitute its curve in the pV-plot for the upper side of the Carnot cycle (see Figure 16) and use the rest of the Carnot cycle to return the system to its starting point. But the entropy at the point A is the same as at D, and at B, the same as at C, because no heat is transferred on the curves DA and BC. Hence the entropy difference between A and B is the same as between D and C, which is Q/T_2. But Q has to be negative, because a positive value of Q would imply that heat is extracted from a reservoir and converted into work with no other changes, which would contradict Kelvin's postulate. Therefore the entropy at D is smaller than at C, and hence that at A is also smaller than at B. Thus, in the irreversible process, the entropy has increased. If the system is not in an equilibrium state, we divide it up into smaller pieces that are, and infer the same.

We therefore have come to the conclusion that the second law of thermodynamics implies the "principle of increasing entropy": *the entropy of any isolated system can never decrease.* But there is no prohibition against a decreasing entropy for a system that is not isolated; entropy can flow out of it and into another system. If the two systems together are isolated, however, their total entropy cannot decrease. Every isolated system must inexorably move toward its maximal entropy, if such a maximum exists (and we shall see later that it does). The entropy of the universe, which is presumably an isolated system, can therefore also never decrease. Thus Clausius formulated the first and second laws of thermodynamics in the positive form: "The energy of the world is constant; its entropy increases toward a maximum," which no longer says that the second law forbids something.

We can study the course of events involved in the increase of entropy in more detail as follows: suppose a Carnot engine works between some hot reservoir at temperature T_1 and a very cold one at T_0. Then the heat Q_0 wasted into the cold reservoir, as we have seen, is related to the heat Q extracted from the hot one by $Q_0/T_0 = Q/T_1$, and, by multiplying this equation by T_0, $Q_0 = QT_0/T_1$; therefore, the work W_1 it produces by extracting the heat Q from the hot reservoir is given by $W_1 = Q - QT_0/T_1$. On the other hand, if the same engine works between the temperature T_2, lower than T_1, and the same cold reservoir, the work it can produce by extracting the same quantity of heat is $W_2 = Q - QT_0/T_2$, which is less than W_1 by the amount $W_1 - W_2 = T_0(Q/T_1 - Q/T_2) = T_0S$, where S is the change in entropy in an imagined flow from the first to the second reservoir. In other words, the quantity of work that can be produced by extracting a given amount of heat from a system (with waste heat going into a given much colder reservoir) is lowered when the temperature of the system is lowered. Therefore, as heat flows irreversibly (for example, in a heat-conducting metal rod) from a hot object toward a cold one, that heat has, in a certain sense, been *degraded;* it can no longer produce the same amount of mechanical energy, and the increase of entropy in the irreversible process is a measure of that degradation.

So every irreversible process makes more and more heat unavailable for work. It is not that the heat energy is lost, but it exists in a less useful form. The hotter the reservoir in which heat is stored, the more useful that heat is, and the continual increase in entropy of the universe means the heat energy in it becomes continually debased by being stored at a lower temperature. If the entropy ever reaches its maximum, the universe has to be in a state of equilibrium, with no further irreversible processes possible. Since only reversible processes can take place in it, it must be at a uniform temperature, and no solid objects subject to friction can be present: this is called the "heat death of the universe." Everything is not necessarily at a standstill, but nothing of any significance can happen any more.

The Need for an Explanation

I am sure it has not escaped the reader's attention that the laws of thermodynamics have a somewhat different "feel" from those of mechanics, which we discussed in the last chapter. As Bridgman put it, they "smell more of their human origin." First of all, these fundamental laws, while of great generality and wide scope, do not have the predictive power of Newton's second law, which allows us to compute the detailed motion of arbitrary objects if we know the forces acting on them. Thermodynamics remains much closer to the observed phenomena themselves; that is why I referred to it earlier as "phenomenological." These phenomena, and the postulates themselves, seem to cry out for an explanation that thermodynamics does not really furnish.

Furthermore, there is a fundamental difference between Newtonian mechanics and the second law of thermodynamics, because Newton's second law of motion and its consequences for the trajectories of particles do not define a sense or arrow of time. A movie taken of the motion of the planets around the sun could be run in either direction, and both depicted scenarios would be equally acceptable. The principle of increasing entropy, on the other hand, defines a unique direction of time, namely that direction in which the entropy increases. If a film of an isolated irreversible process is run backwards, it can be recognized as being "wrong." (This in spite of the fact that thermodynamics itself does not really describe processes as functions of the time.) A video tape of the motion of an idealized set of balls on a billiard table can be run backwards and we could not tell the difference, but real billiard balls have friction and will eventually come to rest; if we ran a tape of those motions backwards, it would be recognized as being viewed in the wrong direction.

The needed explanation of the principles of thermodynamics in terms of more fundamental laws was supplied in the second half of the nineteenth century by three physicists: James Clark Maxwell, Ludwig Boltzmann, and Josiah Willard Gibbs. Maxwell, a Scot, who was a professor at Cambridge, made seminal contributions not only in this area but in a number of others, most notably in the theory of electromagnetism; we shall meet him again in Chapter 4. The framework within which the new ideas of all three men were based was the atomic theory, also called at the time the *mechanistic theory,* according to which all matter is made up of particles that follow Newton's laws and which are so small that no one had ever seen them. At the end of the nineteenth and the beginning of the twentieth century, this was still a controversial theory, even among many reputable scientists who reacted against the seemingly inexorable advance of Laplacean mechanistic determinism. (The atomic theory did not become universally accepted until after Einstein's explanation, in 1905, of the random motion of small particles such as flower pollens suspended in a liquid, which had been observed under a microscope in 1827 by the

Scottish botanist Robert Brown: the cause of the *Brownian motion* of the pollens, Einstein reasoned, is the bombardment of them, from various directions, by the particulate constituents of the liquid. Thus the action of the molecules of the liquid was made directly palpable.)

Boltzmann, an Austrian professor at the University of Vienna, was subjected to sometimes vicious attacks for propagating the mechanistic theory; and just before the tide was beginning to turn in favor of accepting the existence of molecules, he ended his life by suicide— whether out of despondency over these attacks has not been definitely established. Gibbs, who coined the now universally adopted name *statistical mechanics* for the new science he very much enriched, led a more sheltered existence. Except for three years of study in Europe, he spent his life as a professor at Yale in the same house in New Haven in which he had grown up as the son of a Yale professor, rarely traveling. Though he published his work in the obscure *Transactions of the Connecticut Academy of Arts and Sciences,* he became the first American theoretical physicist of high international stature; Europe knew of him, because he sent copies of his scientific papers to prominent scientists there, and Maxwell's support especially advanced his reputation.

Statistical Mechanics

Maxwell's interest in the kinetic theory of gases, according to which a gas consists of fast-moving molecules or atoms, had been stimulated by the work of Clausius. What particularly appealed to Maxwell was an argument by which Clausius had demolished an important objection to the kinetic theory. If the gas molecules moved as fast as Clausius and others had claimed (at least about 1,000 feet per second), why does the odor of a smelly gas diffuse in a room as slowly as it does? Clausius's answered that the gas molecules do not all travel at the same speed, and they continually collide with one another. Therefore, their paths from one part of the room to another are not straight but zigzag and meandering. In fact, Clausius had calculated the probability of such collisions and the average distance a molecule travels between successive encounters. Very much extending this idea, Maxwell introduced the novel method of *statistics* for the description of physical properties, and as a first result, he showed how the velocities of the various individual gas molecules vary when the system is in thermal equilibrium.

Suppose we think of the state of one of the many gas molecules in a given container as a point in its six-dimensional phase space, as we did in Chapter 2 (remember, three dimensions for its position and three for its momentum or velocity), and we divide that phase space up into little cells of equal volumes. What is the probability of finding the molecule in any specified cell? If we regard the molecules in the gas as independent of one another (except for the fact that they frequently collide and therefore continually exchange some of their energies), then this probability tells us at the same time how many of the molecules in the gas have positions and velocities that fall into that same cell, just as the statement that the probability is one sixth for one die to show 3 is equivalent to the assertion that among a very large number of randomly thrown dice, it is almost certain that one sixth of them will show 3.

Maxwell proved that if the gas is in thermal equilibrium at the temperature T, the number of gas molecules that have the speed v is a function, now called a *Maxwellian distribution* (see Figure 17), that vanishes at $v = 0$, peaks at the velocity $\sqrt{2kT/m}$, and rapidly falls off to zero for large speeds, m being the mass of the molecule and k a universal number now called *Boltzmann's constant;* the average speed of the molecules is given by $\sqrt{8kT/\pi m}$. (For nitrogen at 0°C the most probable speed is 401 meters per second, and the mean speed is 453 meters per second.) Experimental measurements have confirmed these values. The average or most probable speed of the molecules, then, is proportional to the square root of the temperature and inversely proportional to the square root of their mass. Gases made up of lighter molecules are more volatile: at the same temperature their molecules move faster. What is more, the *dispersion*—the spread-out-ness—of the velocity distribution also rises with the temperature. At higher temperature, not only is the average molecular velocity larger, but a larger fraction of the molecules have an even higher velocity than the average. At very low temperature, most molecules move very slowly, and their speeds are almost all the same.

The general idea behind the use of the statistical method for the description of the way a system of many degrees of freedom, such as a gas with many molecules, behaves is that it is practically impossible to know its exact state at any one time, as would be necessary in order to predict its future behavior according to the laws of mechanics. This should not really concern us. What interests us is the behavior of the gas as a whole, described not in terms of the precise locations and velocities of all its constituents but in terms of its thermodynamic variables—pressure, volume, and temperature. Since

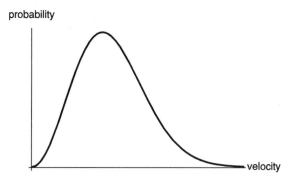

probability

velocity

FIGURE 17. A Maxwellian distribution of velocities.

two gases made up of the same kinds of molecules and having the same volumes, pressures, and temperatures can be expected to differ considerably in the respective locations and velocities of their individual molecules, it is clear that what we want to know are always *averages* over many systems that are made up of the same kinds of constituents.

In order to define such averages, we have to imagine a very large number of systems consisting of the same kinds of molecules; Gibbs used the name *ensemble* for such a collection of identical systems. We can then take *ensemble averages* of quantities that are functions of the positions and velocities of the molecules, just as we can ascertain the mean weight of inhabitants of New York. Our ultimate purpose is to use these mean values and apply them to every individual system. Since we can also calculate the typical fluctuations away from the mean in an ensemble, we have a good idea of what to expect in individual cases, especially because it turns out that in most instances of any interest these average fluctuations are extremely small. Nevertheless, we are dealing with statistics, and on rare occasions exceptions are to be expected.

Recall now our discussion of the phase space of a mechanical system in Chapter 2. For a gas containing 10^{23} molecules, this space has 6×10^{23} dimensions, three for the position of each molecule and three for its momentum. The state of the gas, as far as Newtonian mechanics is concerned, is completely specified by a point in this space. (Note that in contrast to our discussion of the Maxwell distribution, we are now fixing our attention not on the phase space of an individual molecule but on that of the entire gas.) An ensemble of identical gases is therefore specified by a large number of such points on the same energy surface (see Chapter 2) of the same phase space, and the average value of any function of interest is calculated by averaging the values of that function on all these points. But how shall we distribute the points? Where should they be clumped and where thinned out? If we imagine the phase space divided up into small cells of equal volume, one cell having more system-points in it than another, there is a correspondingly higher *a priori* probability for picking a system in the first than in the second. (If you distribute numbered pebbles over a chess board and allow a roulette wheel to pick out one pebble, the probability of picking one on a crowded square is higher than choosing one on an uncrowded square.) Therefore, to avoid bias we distribute the system points evenly and calculate averages on the basis of such *equal a priori probablities.*

There is another argument in favor of such a choice. In our discussion of nonintegrable systems in Chapter 2, I mentioned that most such systems are *ergodic,* meaning that in the course of a long period they will spend equal fractions of time in each of our imagined cells in phase space. Therefore, it is true for almost every such system (but not necessarily all) that if, in the course of its development over some long stretch of time, we choose a specific moment at random to examine it, there is an equal probability for finding it in any cell. This implies that the ensemble averages of arbitrary functions are, for almost all individual systems, equal to the *time averages* of the same functions, and in many cases, such averages over time for individual systems are of primary interest.

Clearly, if our aim is to understand the underlying reasons for the first and second laws of thermodynamics, that aim is immediately achieved for the first law by the adoption of the mechanistic theory. In this theory, heat is described as a random motion of molecules satisfying the energy-conserving Newtonian laws; it therefore follows that for any system as a whole, energy is conserved. As I indicated earlier, the first law becomes, in the context of mechanics, redundant. The same is not true of the second law, which in the context of mechanics becomes even more puzzling. Newtonian mechanics has no arrow of time. Where, then, does that arrow in thermodynamics come from?

Entropy and the Second Law in Statistical Mechanics

To answer that question, let us first go back to the six-dimensional phase space of an individual molecule in our system, but let us picture all the molecules represented by dots in that same phase space. Again we divide the space into a large number of small cells of equal volume and assign a number to each cell, thereby enumerating them. Suppose that the state of the gas at a certain time is such that n_1 molecules are located in cell #1, n_2 are in cell #2, n_3 in cell #3, and so on. It was Boltzmann's idea to define the H-function

$$H = n_1 \log n_1 + n_2 \log n_2 + n_3 \log n_3 + \ldots$$

(not the Hamiltonian, which was also called H in Chapter 2). (For those readers who need a refresher course in logarithms, see the box on page 74.) As time goes on, of course, the numbers of molecules in the various cells change and H is not constant but varies. Boltzmann then proved, taking all possible motions of the molecules and collisions between them into account, that in the course of time the function H can *never be expected to increase*. In fact, we must always expect it to decrease until it reaches its minimum value (at which point the system is in equilibrium, with all cells of the same energy equally populated, and those of different energies populated according to a Maxwellian distribution). What is more, the higher the value of the function H rises above its minimum, the faster it tends to decrease. This is called *Boltzmann's H-theorem*, his crowning achievement.

Another way of expressing the function H, or a function H that differs from it by the addition of and multiplication by irrelevant constants, is to realize that if there are N molecules altogether, the ratio n_1/N is the probability p_1 of finding a particular one of them in the first cell, n_2/N the probability p_2 of finding it in the second cell, and so on. Therefore the H-theorem applies equally well to the function defined by

$$\mathbf{H} = p_1 \log p_1 + p_2 \log p_2 + \ldots$$

A third way of expressing it is in terms of the probability P of finding the whole gas in any particular state, as specified by having n_1 of its molecules in cell #1, n_2 in cell #2, and so on, rather than the probabilities of one of the molecules being in the various cells, and that is

$$\mathcal{H} = -\log P.$$

This is conceptually the simplest function to deal with.

In spite of the fact that our original definition of the entropy in the context of thermodynamics looks quite different, apart from an arbitrary and irrelevant additive constant, Boltzmann's H can be shown to be proportional to the negative of the entropy. Therefore, the H-theorem is equivalent to the principle of increasing entropy, and its proof amounts to a proof of the second law of thermodynamics from the assumptions of statistical mechanics. But, again, what does this proof really tell us about the validity of the second law, and how did Boltzmann manage to wrest an arrow of time from the arrowless structure of mechanics?*

The main thing to remember is that the H-theorem is a *statistical statement.* It does not assert that H for any given system *must always* decrease; it asserts only that it will do so *on the average,* or that its *most probable* change is a decrease. And the crucial assumption is that we are dealing with a mechanical system whose microscopic state is not precisely specified, as is appropriate for thermodynamics. That lack of precise specification is evident in our choice of dividing the molecular phase space into cells of finite size; we were unconcerned about the exact location of the point representing the molecule within a cell, nor did we care which particular molecule was in it.

Let us look at a simple example: a container is divided in half by a wall, and one compartment is filled with a gas while the other one is empty. When we open a hole in the wall, we know from experience that some of the gas will flow into the empty compartment until the pressure and temperature in both halves are equal. That this is also what the H-theorem implies we can see in the following way: if we imagine dividing the phase space of the gas in the container (with the hole open) into little cells of equal volume, with each cell defining the positions and velocities of all the molecules within tolerances determined by its size, there are many more cells corresponding to the molecules being equally distributed among both compartments than there are corresponding to one half of the container being empty, because there are many more ways of realizing that state by distributing different molecules between the two compartments. Therefore, the probability for the molecules to occupy both compartments in equal numbers is much higher than for them to occupy only half, with the other half evacuated, just as, if you toss 1,000 coins, you are

---

*As we shall discuss later in this book, classical Newtonian mechanics has, strictly speaking, been superseded by quantum mechanics. However, there is also *quantum statistical mechanics,* in which quantum mechanics replaces classical mechanics as the underlying description of molecular behavior. Its results are, in general, similar to those of ordinary statistical mechanics, and in some cases more precise. Since quantum mechanics, too, lacks an arrow of time, the puzzle arises in that context as well.

TIME'S ARROW

much more likely to end up with 500 heads and 500 tails than with no heads at all; what is more, the larger the number of coins, the truer this will be. Since the number of molecules in the gas is enormous, the probability of finding one half empty is tiny compared with the probability of seeing them distributed half and half between the two compartments. By the formula above, this implies that $\mathcal{H}$ is lower and the entropy higher if the gas fills both halves; the H-theorem asserts that when you open the hole, the gas will flow into the empty compartment because that will decrease the value of H.

Similarly for two rooms in one of which the air is hot and in the other cold. When you open the door between them, the temperature will equalize, because if we divide the big phase space of the air molecules in the two rooms into little cells, there are many more cells that correspond to an equal velocity distribution of the molecules in both rooms than there are with the faster ones in the first room and the slower ones in the second. Therefore, the probability for a point in the phase space to be found in a cell corresponding to equal speed distributions is much higher than in one that corresponds to a very unequal division of speeds between the two rooms. Thus the H-theorem explains why the heat in the two rooms equalizes, in accordance with our experience and in agreement with the second law of thermodynamics.

But how can we reconcile these statements with the fact that, if we imagine taking a videotape of the molecules rushing from chamber A through the hole into the empty chamber B and filling it, running that videotape backwards produces a scenario that is equally allowed by the laws of mechanics: now all the molecules in B escape through the hole and fill up A, leaving B empty? Similarly for the two rooms after they have settled down to equal temperatures; running a movie of the motions of all the molecules backward shows a chain of events in which one room ends up hot and the other cold. The laws of mechanics say these weird scenarios are not only allowed but are in some sense just as likely to occur as the one we actually see, because for every motion going one way there is an equally possible one going the other way. However, the H-theorem claims they are "forbidden," and yet the H-theorem was supposedly derived from Newton's laws.

Another objection arises from Poincaré's recurrence theorem, which you may recall from Chapter 2: every mechanical system will eventually return to the vicinity of its initial state. How, then, can its entropy be always increasing, or even be rising on the average? There appears to be an irreconcilable conflict.

The function $y = f(x) = \log x$, defined for all positive numbers x, has the important property that for any two positive numbers a and b,

$$\log ab = \log a + \log b;$$

for example, $\log 35 = \log 7 + \log 5$. Except for an arbitrary multiplicative constant, this property uniquely defines the function. The multiplicative constant can be fixed by requiring that for some specified number c, $\log c = 1$; the constant c is then called the *base* of the logarithm. The most commonly used bases are 10, 2, and the number $e = 2.718 \ldots$; if e is used as a base, the function is called the *natural logarithm,* also written "ln." For negative values of its argument, the logarithm takes on imaginary values; the natural logarithm has the remarkable property that $\ln(-1) = i\pi$, where $i = \sqrt{-1}$ and π is the number you know as the ratio of the circumference of the circle and its diameter.

The function $\log x$ increases monotonely with increasing x; it is negative for x smaller than 1, passes through zero for $x = 1$, and then increases without bounds as x grows.

Before the advent of the pocket calculator, logarithms served an important practical purpose, because the fact that the logarithm of a product of two numbers equals the sum of their logarithms very much simplifies the job of multiplying two large numbers. As a result, tables of logarithms were commonly used in schools, and the slide rule, which is based on the use of logarithms, was an indispensable tool of engineers. Now that cheap pocket calculators are ubiquitous, tables of logarithms and slide rules have gone the way of the typewriter.

The inverse of the natural logarithm is the *exponential* or natural growth function in the sense that if $y = \ln x$ then $x = e^y$, which also plays an important role in many areas of physics and mathematics. For example, the Maxwellian distribution of the velocities of the molecules of a gas at the temperature T, plotted in Figure 17, is given by the function $v^2 e^{mv^2/2kT}$. The fact that $\ln(-1) = i\pi$ means that $e^{i\pi} = -1$, a miraculous looking formula first given by the Swiss mathematician Leonhard Euler.

Reconciliation

In order to understand what is going on here, we have to talk about the *fluctuations* of H above its minimal value, or of the entropy below its maximum. Each cell in the big phase space of the gas defines a distribution of the molecules and their velocites, and at a given moment that distribution may be very near or very far away from equilibrium; such distances from equilibrium we call fluctuations. The state of a given system (gas) is represented by a point in its phase space, and as time goes on that point moves from one cell to another. The vast majority of the cells define states near equilibrium, small fluctuations that are macroscopically not even noticeable; and the farther away from equilibrium, the rarer they are. Occasionally, there are more pronounced fluctuations, but the larger they are, the more rarely they occur in the course of the history of the system. That history, completely determined by the laws of mechanics, shows no intrinsic difference between the two directions of the time; forward is as valid as backward. Furthermore, even a very large fluctuation will eventually (after an eon longer than the age of the universe) recur, in agreement with Poincaré's theorem. However, when we specify an initial condition such as "all molecules in one half of the container" or "the molecules in the first room move faster, on the average, than those in the second," we are specifying a state that corresponds to an extremely large fluctuation which would, on its own, occur only once in an enormously long time interval. But we did not wait for it to occur by itself; we produced it by external intervention. If the system is then left to its own devices, it will naturally evolve toward states that differ from equilibrium by much smaller fluctuations, because in the natural course of events a very large fluctuation almost certainly sits in the middle of a vast sea of smaller ones; this explains why H will decrease and the entropy will increase. At the same time it is also true that if we were to follow the system from the chosen starting point backwards, assuming that it was not set up artificially but came to that point naturally, H would decrease in that backward direction too! The time symmetry is thus maintained.

The arrow of time in the second law of thermodynamics, therefore, orginates from two specifications: that the inital condition differs from equilibrium by a very large, *macroscopically* detectable amount, and that we are asking for the development of the system *after* a special condition created by external means rather than by the natural development of the system in isolation. If we were to ask the "unnatural" question: how does the air on its own develop in such a way that it ends up with one room hot and the other cold? the answer would show a development of *decreasing* entropy. But our normal preoccupation with causal chains of events makes this "unnatural" question uninteresting and irrelevant. We therefore have to conclude that the arrow of time introduced by thermodynamics via its second law is not independent of but, in fact, determined by the law of cause and effect (which we shall discuss again in more detail later in this book).

That the second element just mentioned, the *macroscopic* nature of the deviation from equilibrium implied by our initial conditions for a system containing very many *microscopic* particles, is also essential for the functioning of the second law is made clear by a thought experiment proposed by Maxwell. Suppose again we have a vessel filled with gas in equilibrium, separated by a wall from an empty second vessel. The wall has a hole shut by a door that is operated by a "very observant and neat-fingered being," as Maxwell called him, who watches each of the molecules near him individually. Whenever a fast molecule comes rushing toward the door, he quickly opens it to let the molecule into the second vessel; for slow molecules he provides no such escape; after a while one of the vessels will be filled with hot, and the other with cold gas. Since the opening and closing of the door can presumably be accomplished without the expenditure of any significant amount of energy, Maxwell's demon has therefore managed to circumvent the second law of thermodynamics: the temperature difference between the two containers can be used to produce mechanical work, and we have a *perpetuum mobile* of the second kind. Where is the flaw?

One problem with Maxwell's demon is that in order for the imp to observe the molecules, he needs an instrument like a flashlight for their detection. But this requires the expenditure of energy, in the form of radiation to be precise, which would also have to be in thermal equilibrium with his gas surroundings, and hence in a random distribution. That, however, would make it impossible for him to detect the fast molecules in any reliable way.

The more important point to note is this: Maxwell's demon shows that the division between our macroscopic world and the microscopic substratum that underlies it is essential for the existence of the second law of thermodynamics. If that division could be breached in a controllable manner, there would be no second law. This thesis is also confirmed by the argument I presented earlier, that the initial condition for a state that develops with increasing entropy must differ from equilibrium by a macroscopically detectable amount; it must not be a tiny fluctuation whose existence could be observed only by microscopic means.

Other Uses of the Entropy Concept

The principle of increasing entropy is sometimes paraphrased by saying that the "disorder" of the world keeps on growing in the course of time. An ancient temple, left alone and unsupervised, will, according to the second law, inevitably decay and eventually become a heap of rubble; a highway, left unrepaired, must return to the natural state of the soil. Why are such predictions legitimate consequences of the second law of thermodynamics?

Whatever, in any context, we mean by the concept of "order" has to involve as an essential element the idea that what is ordered is the opposite of "random." If we praise the artistry of a painting, one thing we assume is that it was not created "by chance"; an engine whose efficiency we admire cannot have been thrown together by happenstance. In other words, a highly ordered system is, by definition, a very *improbable* one. A heap of rubble, on the other hand, is a very probable state of affairs. Why? By its very definition, a highly ordered system is one in which even a tiny flaw diminishes its orderliness. A nick in a piece of sculpture offends the eye; a performer's small error in a passage of music grates on the ear; and a scratch on a new automobile sends the owner to the dealer with complaints. Therefore, among the myriads of possible arrangements of a heap of steel, it is only a tiny number that take the form of a Cadillac; the vast majority are scrap, because we make no distinction between one kind of scrap and another. A scratch or dent on a piece of scrap metal is not even noticeable: a very large number of states are called "scrap" but only a tiny number form a Rolls Royce. Therefore, the Rolls Royce is a very improbable arrangement, and the scrap a very probable one. This is the reason why the principle of increasing entropy predicts that an automobile, left alone, will eventually turn to scrap, and not the other way around. (Note, however, that if we single out one particular shape of the scrap and call it a work of art, it automatically becomes a state of order because it is not just any old mount of rubble, but a configuration of a special shape. Some modern pieces of art, alas, seem to undermine this particular argument.) "Disorder" is the label we attach to a very large number of states, while the label "order" is applied only to a very small fraction. This makes order highly improbable and disorder likely, so that order has a low and disorder a high entropy.

A quite different context in which entropy is used for similar reasons is that of information theory. The amount of information contained in a message of a given length is very small if the arrangement of the elements of which it is composed is almost chaotic; if it is completely random, there is no information in it at all. On the other hand, if the arrangement is highly ordered, that is, *a priori* very improbable, then it carries much information. So the amount of information varies with the probability of the arrangement. Or to put it another way: a message whose content has an *a priori* probability equal to one carries no information, but a message that says something very unlikely contains a lot of information. It has therefore become customary to use the negative of the entropy (sometimes called "negentropy," which is, of course, the same as Boltzmann's H) as a measure of the information content of a message, an idea found to be very useful. In other contexts, such as literature or philosophy, the concept of entropy is used figuratively, and sometimes without much understanding or justification.

In these last two chapters we took the forces that act on the constituents of a physical system as given and did not examine the way in which they acted. Let us now look more closely at the nature of these forces.

4

FORCES ACTING
THROUGH SPACE

For Descartes, as for the ancient Greeks, a force could be transmitted only by direct contact. No body could exert a force on another without touching it. The same idea also underlay the natural philosophies of Thomas Hobbes and Pierre Gassendi, the main advocate of atomism in seventeenth-century France. It was, in fact, the basis of the mechanistic philosophy that pervaded the time. Imagine, then, their reaction, and the reaction of the followers of this philosophy, to the idea proposed by Isaac Newton that what made the earth revolve around the sun and the moon around the earth was the universal force of gravity, acting at enormous distances through empty space. To make the apple fall was one thing; after all, there was air between the earth and the apple. But how could the earth be attracted to the sun when there was no intervening matter?

It was therefore no great surprise that Newton's theory of universal gravitation had a very hard time being generally accepted. To be sure, there were philosophers who circumvented such problems by positing the idea of a universal *plenum* or *ether* that filled all space. Descartes thought of this ether as consisting of particles that were in constant swirling motion, forming a perpetual succession of *vortices* which transmitted the forces. In his youth, Newton, too, was a firm believer in such an all-pervading fluid. Indeed, he argued, to suppose "that one body may act upon another at a distance through a vacuum, without the mediation of anything else, . . . is to me so great an absurdity, that I believe no man, who has in philosophical matters a faculty for thinking, can ever fall into."

Later, however, he reversed himself in favor of what came to be called *action at a distance.* This concept requires no intervening matter to transmit forces from one point in space to another even over large separations. The force exerted by the sun on a planet acts instantaneously and directly, leaping, so to speak, from one to the other. He came to apply this idea to virtually all natural phenomena. "Yet what we have said about these forces will appear less contrary to reason," he now maintained, "if one considers that the parts of bodies certainly do cohere, and that distant particles can be compelled towards one another by the same causes by which they cohere, and that I do not define the manner of attraction, but speaking in ordinary terms call all forces attractive by which bodies are impelled towards each other, come together and cohere, whatever the causes be."

The hostile reaction he encountered to such a perverse idea was exemplified by that of the Dutch physicist Christiaan Huygens: "I don't care that he's not a Cartesian as long as he doesn't serve us up conjectures such as attraction." Leibniz regarded the Newtonian views as a return to the discredited scholastic concept of *occult qualities.* In his *Lettres philosophiques,* Voltaire described the difference in philosophical atmosphere that had developed by 1730 between England and the Continent: "A Frenchman who arrives in London will find philosophy, like everything else, very much changed there. He had left the world a *plenum,* and now he finds it a *vacuum.*" In a private letter he explained, "It is the language used, and not the thing in itself, that irritates the human mind. If Newton had not used the word *attraction* in his admirable philosophy, everyone in our Academy would have opened his eyes to the light; but unfortunately he used in London a word to which an idea of ridicule was attached in Paris; and on that alone he was judged adversely, with a rashness which some day will be regarded as doing very little honour to his opponents."

Newton could not leave this mysterious transmission of a force through empty space unexamined; he felt compelled to find an "explanation" for it. When he discussed, among other things, the question "What is there in places empty of Matter, and whence is it that

FORCES ACTING THROUGH SPACE

the sun and planets gravitate towards one another, without dense matter between them?" with the mathematician David Gregory, the latter recorded the event as follows:

His doubt was whether he should put the last Quaere thus. *What the space that is empty of body is filled with.* The plain truth is, that he believes God to be omnipresent in the literal sense. And that as we are sensible of Objects when their Images are brought home within the brain, so God must be sensible of every thing, being intimately present with every thing: for he supposes that as God is present in space where there is no body, he is present in space where a body is also present. But if this way of proposing this his notion be too bold, he thinks of doing it thus. *What Cause did the Ancients assign of Gravity.* He believes that they reckoned God the Cause of it, nothing else, that is no body being the cause; since every body is heavy.

After Newton, the concept of action at a distance gradually became the dominant natural philosophy, though it did not manage to win over everyone.

Faraday's Lines of Force

Let us now move ahead to the middle of the nineteenth century. Michael Faraday, the son of a blacksmith and largely self-educated, was the experimental scientist *par excellence*. Despite the fact that he knew little mathematics, which gave him the undeserved reputation of being opposed to its use in physics, he developed a strong theoretical framework that guided his experiments and led him to many discoveries; and that framework did not include the idea of action at a distance. Because action at a distance was strongly believed in by almost all scientists of the day, he proceeded very cautiously, but finally decided that his experiments on electrostatic induction, as he called it, could not be interpreted by that concept.

Influenced by his early experiments with magnets, Faraday had developed the idea of *lines of force* which at every point in space show the direction of the force exerted on another magnet, or on an electric charge, as the case may be. While their direction indicates the direction of the force at a given point, the density of the lines indicates the strength of the force. As they spread, they grow farther and farther apart and the force gets weaker and weaker. A vivid embodiment of these lines of force can be seen in the manner in which iron filings on a piece of paper arrange themselves in the presence of a

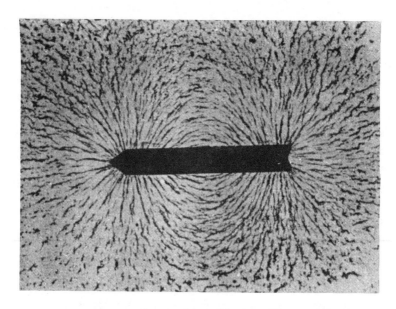

FIGURE 18. Pattern formed by iron filings in the neighborhood of a bar magnet.

magnet (see Figure 18). In the case of electrostatic forces, there is no similarly easy pictorial demonstation. However, Faraday argued that if the influence of an electric charge on a conductor, and vice versa, is exerted by action at a distance, it will not depend on any condition of the intervening space. Furthermore, there is no reason for these lines of force—which he envisaged as, in a manner of speaking, guiding the force—to be other than *straight*. But his experiments found them to be curved! "As argument against the received theory of induction," he wrote in 1837, when publishing the results of his experiments with metal planes and spheres,

I cannot see how the preceding results can be avoided. The effects are clearly inductive effects produced by electricity . . . and this induction is exerted in lines of force which, though in many experiments they may be straight, are here curved more or less according to circumstances. I use the term *line of inductive force* merely as a temporary conventional mode of expressing the direction of the power of induction; and . . . it is curious to see how, when certain lines have terminated on the under surface and edge of the metal, those which were before lateral to them *expand and open out from each other,* some bending round and terminating their action on the upper surface of the hemisphere, and others meeting, as it were, above in their progress

FORCES ACTING THROUGH SPACE

outward . . . All this appears to me to prove that the whole action is one of contiguous particles . . . So that in whatever way I view it, and with great suspicion of the influence of favorite notions over myself, I cannot perceive how the ordinary theory applied to explain induction can be a correct representation of that great natural principle of electrical action.

(See Figure 19.) Faraday considered his lines of force to be quite different from the force of gravity and more analogous to the action of magnets. Notice that for an explanation, he returned to the earlier Cartesian notion of forces being transmitted by contiguous particles.

Fifteen years later, many new experiments had altered the picture, and he changed his ideas on the nature of the lines of force. The French physicist André Marie Ampère had discovered that wires carrying electric currents produced magnetic forces in the same way as magnets, and these forces were by no means always directed along a straight line from their source. On the other hand, Faraday had found that a magnet that moved in the vicinity of a wire produced a current in it, an "induction" which is now known as *Faraday's law*. By that time he had given up trying to explain these strange effects by the action of contiguous particles and boldly proclaimed them to be there even in empty space, without the need for an intervening ether or *imponderable fluid*. "If [the lines of force] exist," he wrote, "it is not

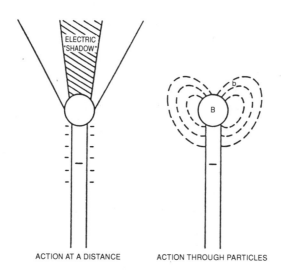

ACTION AT A DISTANCE   ACTION THROUGH PARTICLES

FIGURE 19. Faraday's figures contrasting action at a distance with action through particles.

by a succession of particles as in the case of static electric induction, but by the condition of space free from such material particles. A magnet placed in the middle of the best vacuum we can produce . . . acts as well upon a needle as if it were surrounded by air, water or glass; and therefore these lines exist in such a vacuum as well as where there is matter."

But do these lines of force in space have physical reality, he wondered, or are they merely convenient representations of something else?

With regard to the great point under consideration, it is simply whether the lines of magnetic force have a *physical existence* or not? Such a point may be investigated, perhaps even satisfactorily, without our being able to go into the further questions of how they account for magnetic attraction or repulsion, or even by what condition of space, ether or matter, these lines consist.

He began to believe strongly in the physical existence of the lines or tubes of force, and not only for electric and magnetic phenomena but also for gravitation, where, because of Newton, action at a distance held sway most strongly. All three of these forces were transmitted by a condition of empty space. In other words, he no longer thought of such forces as being exerted directly by one particle or object upon another but as an influence of the first on all surrounding space, with the force on the second being exerted directly by the condition of space at the point of its position.

Much as Faraday was admired, the scientific establishment was not easily persuaded of his theories. The reaction of George Biddell Airy, the Astronomer Royal, was not at all atypical: "When I contemplate gravitation, I contemplate it as a relation between two particles, and not as a relation between one particle (called the attracting particle) and the space in which the other (called the attracted particle) finds itself for the moment."

Faraday's Field

In the face of such reluctance by his fellow scientists, Michael Faraday, the experimenter, originated what is now the dominant theoretical concept in almost every part of physics, namely the idea of the *field*. All fundamental forces of nature are described by fields. Lest this appear to be mere metaphysics and of no real import, let us consider the mathematical aspect of this development. Take, for example, Coulomb's law, named after the French physicist Charles Augustin Coulomb. In its original form, it states that the force between two electric charges equals the product of the charges divided by the square of the distance between them, and it points in the direction of the line that connects the two charges. After the introduction of the field concept, we now say instead that the first charge produces a condition of space everywhere, called the electric field. This field falls off in strength with the inverse square of the distance. If another charge is placed at any point in space, the force exerted upon it is the action of the field at the very point of its position. But that is not all.

Once we accept the idea of the field and we reject the notion of action at a distance, so that all influences go from points in space to contiguous points, it becomes clearly desirable to describe the field itself in terms other than the distance from its source. The field itself ought to be described mathematically in terms of the influence of neighboring points. This is what differential equations are ideally suited for. In this context, however, we need the concepts of *partial derivatives* and *partial differential equations*.

In Chapter 2 we defined the derivative of a function of a single independent variable. What happens if the dependent variable z is a function of two independent variables x and y, $z = f(x,y)$? We may then temporarily hold y fixed and, regarding $f(x,y)$ as a function of the single variable x, form its derivative. This is called the *partial derivative* of f with respect to x, and denoted by $f_x(x,y)$. Since such a procedure may by followed for any value of y, the partial derivative with respect to x is a function of x and y. Similarly we may temporarily hold the variable x fixed and form the derivative of the resulting function of y; this gives us the partial derivative of f with respect to y, namely, $f_y(x,y)$. This set of partial derivatives is also known as the *gradient* of f. (For more on partial derivatives, see the box on the next page.)

Field Equations

A partial differential equation is an equation between the values of the unknown quantity, in this case the field, at infinitesimally neighboring points (in space or time, or both). We consider the ratio of field values at nearby points to the distance between these points in the limit as the points approach one another and their distance ap-

PARTIAL DERIVATIVES AND
PARTIAL DIFFERENTIAL EQUATIONS

It is clear that the idea outlined on the previous page for functions of two variables can be used for functions of any number of independent variables, and it may be repeated to form second partial derivatives. We need not do that in detail. For an example of a partial derivative, we may return to our function that gives the temperature T at an altitude h and the time t, $T = f(h,t)$. Its partial derivative with respect to h, namely, $f_h(h,t)$, tells us the rate at which the temperature changes with altitude at the altitude h and the time t; the partial derivative with repect to t, $f_t(h,t)$, tells us the rate at which the temperature changes with time at the altitude h.

Or we may consider the temperature in a region of space in which a point P is specified by the three coordinates x,y,z where x and y are the horizontal distances of P in feet from a designated point in the east-west and north-south directions respectively and z is its height above the ground: $T = f(x,y,z)$. Then f_z is the rate at which the temperature changes with altitude, and f_x and f_y are the rates at which the temperature changes with the horizontal position of the point P in the two directions east-west and north-south. Thus, if we know not only the temperature T at a point P at the height $z = h$ but also its partial derivative with respect to z, we can say that at any other altitude $h + a$ near h, the temperature must be $T \simeq f(P) + af_z(P)$ to a good approximation that gets better the smaller a is; similarly for shifts in the horizontal direction. (Remember that $\simeq$ means "approximately equal.") So knowing not only the temperature at a point P but also its partial derivatives at P (known as its *gradient*) gives us information about the temperature in the neighborhood of P.

A function of several independent variables may satisfy a differential equation that contains partial derivatives with respect to these variables; this is called a *partial differential equation* (in contrast to an *ordinary* differential equation if there is only one independent variable). Again, such an equation has infinitely many solutions, and in order to solve it uniquely it must be supplemented by boundary conditions, or by a combination of boundary and initial conditions if one of the variables is the time.

proaches zero. In this manner we form the derivative or the gradient of the function. To describe the field by *field equations,* which are differential equations, is therefore to say that the real cause of the field at any point is not directly the gravitating body, the electric charge, or the magnet far away but the field itself in an infinitesimal neighborhood, now and at the preceding moment. If we look at nature in this way, all the laws that describe the action of gravitating bodies, magnets, and electric charges and currents should be formulated as differential equations for the corresponding fields.

Differential equations suitable for the electrostatic field had already been set up and investigated by the mathematicians Laplace and Siméon-Denis Poisson. About thirty years after Faraday, James Clark Maxwell in one stroke not only unified the apparently disparate electric and magnetic fields but he performed for the newly combined *electromagnetic field* exactly the same task Laplace and Poisson had for static electric fields alone.

The Maxwell equations completely describe all possible electromagnetic fields, whether static or changing with time, by means of a set of differential equations. Such equations, however, have many solutions, and the particular solution needed is singled out by boundary conditions that describe the contingent aspects of the situation. The electric field of a given collection of point charges will always satisfy the same differential equation. However, if we want to know the field produced by these charges when they are placed inside a metal box in the shape of a cube or in the shape of a sphere, then the boundary conditions in the two cases are different and the unique solution of the field equations with these boundary conditions is the electric field found in that specific physical situation.

With the Maxwell equations and the many consequences they entail, the idea of the electromagnetic field took on a life of its own. By the time of Maxwell, the concept of energy and the law of its conservation were well accepted. It was therefore natural to ask: If electric circuits, say, exert forces upon one another, then what happens to the energy that is added when work is done to move one of them? The total energy is conserved only if we recognize that there is a certain amount of energy stored in the field itself. In other words, there is energy even in the vacuum when it contains electric and magnetic fields. The forces exerted by the field are no longer primary but instead become the manifestations of a *stress* in the vacuum, just as the forces transmitted by a solid are describable in terms of a stress in that solid. What is more, the field even carries *momentum.* Just as any moving material body has both energy and momentum, so does the electromagnetic field, and in any interaction between it and a physical object, the total energy and the total momentum are both conserved. This is the origin of the concept of the *radiation pressure* exerted by electromagnetic waves on objects upon which they impinge.

The difference between an action-at-a-distance theory and a field theory becomes dramatically apparent when you imagine a single electrically charged particle in empty space. If you believe in action at a distance, the particle sits there, devoid of any evidence that it is electrically charged. It exerts no force. But if you describe the situation in terms of a field, you see the charged object surrounded by an electric field that gets weaker and weaker with distance but nevertheless stretches to infinity. Furthermore, this field carries energy, and the field equations allow you to calculate how much it carries. This is the so-called *self-energy* of a charged particle. The exact amount of this self-energy depends on the model adopted for the particle, its radius, the manner in which the charge is distributed inside it, and so on. (If you imagine it as a *point* particle, then the self-energy is infinite, because the field strength, and hence the energy density of the field, grows without bounds as the distance shrinks to zero.) Every electrically charged particle carries with it, drags around with it, so to speak, this infinitely extended cloud of field.

Light and the Ether

The idea that light is a vibrating, oscillating electromagnetic field in which both the electric and the magnetic fields participate greatly expanded the scope of Maxwell's theory. In fact, light may be regarded as a field that has detached itself from its source altogether and is able to extend itself over indefinite distances without necessarily decreasing in strength. According to Maxwell's equations, it moves with a finite, fixed speed in empty space, and it carries with it both energy and momentum. It then stands to reason that, when a beam of light strikes a mirror, it will exert a force of pressure. We do not ordinarily notice this pressure when we are in the beam even of a very bright light because it is extremely feeble. However, we do recognize it in plans to exploit the wind of photons for space travel propulsion.

If we envisage light as vibration, we might very naturally ask

FORCES ACTING THROUGH SPACE

"What is it exactly that is vibrating?" More generally, if the forces transmitted by the electomagnetic field are the result of a stress, what is it that is being stressed? Here we are again, then, back with the concept of the *ether*: it is the *medium* that is being stressed by the field, and also the medium that vibrates to give rise to a light wave. Elaborate models of the ether were constructed to account for its properties. Some thought it was like an elastic solid, others like an incompressible fluid. Maxwell himself devised a model consisting of vortices with particles that served as idle wheels between them, to assure that all the vortices rotated in the same direction. Most of the physicists of the day participated in the various attempts to find an acceptable model of the ether to account for the properties of the electromagnetic field as described by the Maxwell equations. Mathematicians tried their hands, or minds, at this as well; both Gauss and Riemann made unsuccessful attempts at this construction.

The kind of model that physicists had in mind for the ether was analogous to a medium in which sound waves are propagated. In that case, too, the velocity of sound is fixed for a given medium, as is the velocity of light. If we think in terms of such a model, then we might naturally ask, "What is our velocity with respect to the ether?" and there is a straightforward (but very difficult) experimental procedure to determine the answer. If the velocity of light is fixed with respect to the ether, then measurements of the velocity of light in various directions should give varying results, depending on the direction of motion. In the forward direction (with respect to our motion through the ether) the measured velocity should be the difference between the fixed "velocity of light" and our speed; in the backward direction it should be the sum of the two. In principle, this difference should have been detectable; in practice it was very difficult because our velocity with respect to the ether was presumed to be extremely small compared with the speed of light, which is about 3×10^{10}cm/sec or 186,000 miles per second. The experiment, first done in 1887 by two American physicists, Albert Abraham Michelson and Edward William Morley, and repeated many times since then, produced a negative result. This means that the speed of light, as measured on the earth, is the same in all directions. In spite of attempts to "fix things up," to many physicists this was the death-blow to the concept of the ether, and it ushered in Einstein's theory of relativity, the only physical theory that could successfully account for the constancy of the speed of light (see Chapter 6).

The Special and General Theories of Relativity

From that time on, in order not to conflict with experimental results, all field theories have had to fit into a rigid constraint: they must conform to the *special theory of relativity*. The fields must have the right "transformation properties" from one observer to another who moves relative to the first, and the field equations are required to have the proper form so that they look the same to two such observers. (Chapter 6 will explore these transformation properties in more detail.) The Maxwell equations, it turned out after appropriate reformulation, had exactly such a form; this was no accident because, after all, experiments and thought-experiments with light waves known to be described by Maxwell's equations had led Einstein to his theory. But what about Newton's theory of gravity? The instantaneous action of one celestial body on another by "action at a distance," which that theory assumed, is incompatible with relativity. One of the most fundamental results of the theory of relativity is that no physical action or signal can be propagated with *superluminal* speed, that is, with a speed greater than that of light (again, see Chapter 6). A new formulation of the theory of gravitation was called for.

The outcome of this thought process and other profound ideas by Einstein was the so-called *general theory of relativity*. It is a theory of gravity which supersedes that of Newton, but becomes equal to Newton's for slow velocities and small masses. This reformulation of the theory of gravitation as a field theory took a form that differed somewhat from Maxwell's formulation for the electromagnetic field. Because of the universal character of the gravitational attraction, whose effect is quite independent of any particular properties of the objects involved, Einstein cast the theory in the form of *geometry*. In other words, instead of describing the properties of a gravitational field in space and time, the theory describes the properties of space itself. Gravitating bodies have the effect of distorting the geometry of the space around them. If the path of a light ray appears "bent" near the sun, it is because the geometry of the space near the sun is such that straight lines do not have the properties we associate with them in Euclidean geometry. Large masses, especially when concentrated in relatively small bodies, as they are in certain kinds of celestial objects, produce a strong *curvature of space*. The geometry of such a space is very far from the Euclidean geometry we are all accustomed to and which we learn in school. As we have already discussed in Chapter 1, triangles made up of "straight lines" have angles whose sum does not equal 180°, for example, just as the sum of the angles in a triangle on the surface of the earth is always greater than 180°, depending on its size.

Whether you want to call such an effect on the properties of free space a field or geometry is, of course, just a matter of taste or convenience. What is not simply a mode of speaking, however, are the predictions which differed from those of Newton's theory and which are implied by the general theory of relativity, and these were confirmed by astronomers. A number of such agreements with observations established the validity of the theory. I am not so much concerned here with successful tests as with a certain prediction in principle. Maxwell's equations implied the existence not only of static and slowly varying electric and magnetic fields, but also of electromagnetic *waves*. Some of these were interpreted as light, and others were found later in the form of radio waves, microwaves, infrared rays, x-rays, and gamma rays. Similarly, the general theory of relativity predicts the existence of *gravitational waves*. While indirect evidence for the existence of these waves has been found by astronomers, they are very difficult to detect directly, and a number of experimenters have been and are at the present time looking for them. There was a flurry of excitement in 1960 when Joseph Weber, a physicist at the University of Maryland, thought he had found direct experimental evidence for their existence. The rest of the physics community, however, did not regard the data as sufficiently convincing, and the search for direct observation of these waves goes on, so far without success.

The equations of the general theory of relativity are difficult to solve. Certain special kinds of solutions, however, can be given: if the mass of a body is concentrated in too small a space, which can be expected to happen when a massive star burns out its nuclear fuel and keeps on contracting as its mass falls more and more inward, there comes a time at which its gravitational field grows so strong that nothing, not even light, can ever escape from it. This is what John Wheeler called a *black hole,* and the name stuck. These fascinating objects have not yet been unambiguously identified, but astronomers think that there are many of them in the universe. They have found, in fact, somewhat indirect evidence for their existence, for example, in quasi-stellar objects, the famous *quasars,* and possibly at the center of many galaxies.

Faraday had hoped that the theory of gravitation would turn out to be a field theory, and it did. However, even as Faraday had been dreaming of a single theory that would encompass both the electromagnetic field and that of gravity, so the older Einstein spent many years trying to find such a unified field theory. At his death it still eluded him.

Enter the Quantum

In the meantime, however, another development in the theory of light had taken place. In order to explain the distribution of frequencies in the radiation emitted by a "black body," that is, a body that absorbs all radiation and reflects none, the German physicist Max Planck at the turn of the century had, without explanation or justification, taken the light emitted by it as consisting of discrete packages of energy. Their size was proportional to the frequency of the light. (The constant of proportionality is now known as Planck's constant.) More generally, Einstein in 1905 had invented the concept of light quanta or *photons,* an idea that, among other things, explained the photoelectric effect, in which electrons are emitted from a metal surface when light shines on it (and for which he was awarded the Nobel Prize in 1921). From that time on, it was no longer possible to regard light, as it had been since the time of Huygens, purely as an oscillatory phenomenon. It now had to have also some properties that resemble those of a *particle*. Physicists would have to find a way to adapt the field concept to this entirely new situation.

A period filled with revolutionary ideas followed for the next twenty years. Niels Bohr introduced a new model of the atom in which electrons revolved about the nucleus in stationary orbits without emitting the radiation that classical theory would predict, and Louis de Broglie advanced the idea that not only was light, which had previously been regarded as a wave, now both a wave and a particle, but even electrons, which had certainly been thought of as particles, were now to be considered both as particles and waves. Finally, after much healthy turmoil and confusion, in the mid-1920s Werner Heisenberg, Erwin Schrödinger, and Paul Adrian Dirac invented "quantum mechanics." (For further discussion of quantum mechanics, see Chapter 7.) It then clearly became necessary to rethink the whole structure of electrodynamics and of field theory. Almost as soon as quantum mechanics was born, therefore, the concept of *quantum electrodynamics* was invented.

The technique for handling the electromagnetic field with the new mathematical tools was the same as that employed in mechanics. First of all, the system is described by a *wave function*, the squared magnitude of which determines the probability of finding the system, or particles, in a certain region of space or in a certain state. Second, in order to take into account the fundamental *principle of indeterminacy* enunciated by Heisenberg—according to which such quantities as the position and the momentum of a particle can be simultaneously determined only to within limits whose product must always be greater than Planck's constant—all "dynamical variables," such as position and momentum, have to be replaced by *operators*. Operators are quantities that do not simply multiply the wave function of the system by a constant but change it into a different function. In contrast to ordinary multiplication of numbers, for which $5 \times 3 = 3 \times 5$, the action of such operators may depend on the order in which they are applied; *they may not commute*. For example, if a physical system is subjected to the operation of rotation twice about two different axes, the results of performing the two rotations in a different order are not the same (see Figure 20). Imagine rotating the earth first by 90° towards the west about its north-south axis, taking Quito in Ecuador marked **B** in the figure to **B′**, and then by 90° in a right-handed screw sense about an axis through the equator at the new position **B′** of Quito. The north pole, marked **A**, will end up where Quito used to be, and Quito will be in the middle of the Pacific Ocean. But if you perform the rotations in the opposite order, Quito ends up where the north pole used to be.

To "quantize" the classical equations of motion of a mechanical system means to replace the numerical functions that represent the positions and momenta by operators that have prescribed "commutation relations." That is to say, the difference between applying these operators in the opposite order is prescribed in terms of Planck's constant. All the remarkable results of quantum mechanics follow from this simple prescription.

What had to be done to the equations of the electromagnetic field is the exact analogue of this procedure in mechanics. In other words, the functions which determine the electromagnetic field at every point in space and every moment in time had to be replaced by operators. In order to describe the emission and absorption of photons in this theory, we introduce what are called "creation" and "destruction" operators. If a "creation operator" acts on a state of n photons, it converts it into a state of $n + 1$ photons; if a "destruction operator" acts on such a state, it produces a state of $n - 1$ photons. In this way, the theory combines in a natural manner the oscillation phenomena

FORCES ACTING THROUGH SPACE

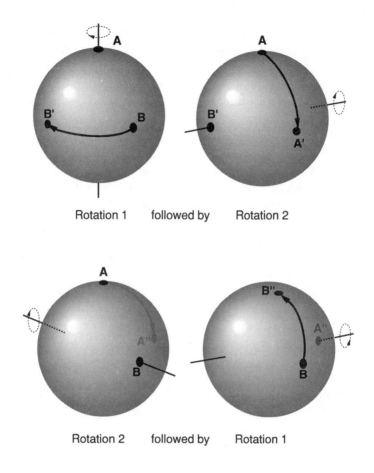

Rotation 1   followed by   Rotation 2

Rotation 2   followed by   Rotation 1

FIGURE 20. Two successive rotations executed in opposite order.

of the field and its particle-like character. However, it is not only the electromagnetic field that is described in this new operator language; the electrically charged particles that are the sources of the fields are themselves described in the same language. There is a "matter field" as well as an electromagnetic field, and the theory has creation and destruction operators for electrons as well as for photons.

The indeterminacy principle of Heisenberg, mentioned earlier, has another aspect to it. Any physical system that is not in a steady state, which means it is in a state that is not constant but varies with time, cannot have a well-defined energy. If we want the system to have an energy that is fixed within certain limits, that system has to be essen-

FORCES ACTING THROUGH SPACE

tially unchanged for a certain length of time. The product of this length of time with the uncertainty of the energy must be at least equal to Planck's constant. Therefore, for short periods of time the law of conservation of energy may be violated, and the shorter that period of time, the more it may be violated.

Combining Quantum Theory with Relativity

At this point, the theory of relativity has to be brought to bear on the field. It was the great contribution of Dirac to have invented a new equation describing electrons in a manner that is in accordance with Einstein's theory of relativity and that also explained for the first time a property of electrons which had been discovered but was thought to be rather mysterious: their *spin*. At the same time the Dirac equation, which physicists greatly admire both for its content and for its beauty, implied the existence of a new particle, which was experimentally discovered shortly thereafter in 1932 by the American physicist Carl D. Anderson, and is now called the *positron*. The positron has the same mass as the electron and the same amount of electric charge, but with the opposite sign. It was the first of a whole new class of particles theoretically predicted and then discovered: each particle in nature has an "antiparticle" of the same mass and certain properties that are equal in magnitude but with the opposite sign of those of the particle. (Some particles are their own antiparticles.)

Putting all these ideas together, we now envisage the process by which particles interact in the following way: first of all, the electromagnetic force exerted by one electron on another is described in the new language as being due to the emission of photons by one electron and their absorption by the other, the way one boy could push another by throwing a ball at him. These photons may not be "real" but "virtual," which means that they lack some of the properties of real photons. In any event, this is how one thinks about the process. Now, photons interact with the "electron field" and can produce a pair consisting of an electron and a positron. (Electric charge is conserved.) This process of pair creation may require a violation of the conservation of energy, because each of these particles has the mass m and therefore, according to the theory of relativity, a minimal energy of mc^2 (where c is the velocity of light). It therefore takes a minimum energy of twice that amount for the photon to create such a pair, and the photon may not have enough of it. But a violation of energy conservation is allowed for a short period of time, after which

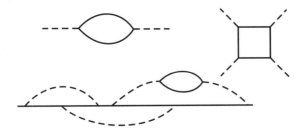

FIGURE 21. Some Feynman diagrams. The dotted lines represent the propagation of photons and the solid lines electrons and positrons.

the members of the pair must come together again and annihilate each other, giving up their energy in the form of another photon. During the short span of the existence of this "virtual" pair of particles, each may, in turn, emit a photon, which, in turn, may create a "virtual" pair for a minute period of time, and so on. (Figure 21 shows several "Feynman diagrams" representing such pair creations and emissions and absorption of photons.) This process is, in principle, unending, but since each of these virtual processes has a rather small probability, we need not take all of them into account.

Quantum Electrodynamics

Mathematical obstacles made the field theory of quantum electrodynamics constructed in this fashion difficult to formulate. The infinite succession of the kind of processes described above led to unacceptable results when one tried to calculate experimentally verifiable predictions. Many of the predicted numbers simply came out infinite. In spite of this, new formulations of the theory by the American physicists Richard Feynman and Julian Schwinger and the Japanese physicist Sin-Itiro Tomonaga led to the calculation of unambiguous, finite results that could be compared with experiments. The agreement between these calculated results and the experimental observations was found to be astonishingly good, in some cases to better than ten decimal places, or one part in ten billion.

Quantum electrodynamics thus became the field theory that has taken the place of the classical Maxwell theory. Maxwell's equations are still valid, but now they are satisfied by operators that represent the electromagnetic field instead of by numerical functions. The picture that emerges, then, of what "empty space" is like is enormously complicated. As we have seen, we have a picture of an endless succession of the creation and subsequent annihilation of pairs of electrically charged particles. This is called the *polarization of the vacuum*. Every charged particle is surrounded by a cloud of photons, each of which in turn produces a cloud of charged pairs that, in effect, alters the net charge as seen from far away. We therefore have to make a distinction between the "bare" charge of a particle and the "renormalized" charge of the "dressed" particle, whose clothes consist of the virtual pairs.

Other Quantum Fields

This complicated picture is further encumbered by the fact that there are two kinds of forces in the microscopic realm that were not at all known at the time of Maxwell and only vaguely at the time of Einstein: the *strong force* is responsible for holding the atomic nucleus together, and the *weak force* gives rise to such effects as radioactive decay of nuclei. These forces, too, can be described by means of field theories. It took another twenty-five years to do this, but the so-called electroweak theory devised by the American physicists Steven Weinberg and Sheldon Glashow and the Pakistani physicist Abdus Salam manages to combine the electromagnetic force and the weak force in one field theory; the theory called *quantum chromodynamics* explains the strong force in terms of another quantum field. Attempts at combining all three of these forces into one field theory are called "grand unified theories," but none of these ambitious attempts has yet been totally successful.

The theories mentioned—electrodynamics, chromodynamics, and the electroweak theory—are all so-called *gauge field theories*. The history of the idea of gauges again goes back to the Maxwell equations. We can write these equations in a much simpler form by introducing what are called *scalar* and *vector potentials,* two functions in terms of which the electromagnetic fields can be expressed. The potentials, however, may be changed in certain specific ways called gauge transformations, without changing the electromagnetic fields themselves—just as the units in which we measure the length of a table may be changed from inches to centimeters without altering the length of the table. Since such gauge transformations of the potentials do not change the electric and magnetic fields, they are considered, at least from a classical point of view, of no physical significance. All experimentally measurable quantities, when expressed in terms of the potentials, should be *gauge invariant,* that is, they should not change if the potentials are subjected to gauge transformations.

In quantum mechanics this gauge transformation has to be combined with a change of the wave function, a *complex number* (see Chapter 1), which you may picture as a point in a plane (see Chapter 2), with the real part plotted along the x-axis and the imaginary part along the y-axis. All observable consequences of the theory depend only on the distance of this point from the center; therefore, a simple rotation of the point about the coordinate origin as the center should have no physically observable effect. This requirement, together with the requirement of invariance under gauge transformations of the electromagnetic potentials, constitutes the quantum mechanical constraint of gauge invariance, to which all observable quantities are subject.

Gauge Fields and Strings

An enormously fruitful idea of the Chinese-American physicist Chen Ning Yang (who had won the Nobel Prize for something altogether different, which we shall discuss in Chapter 10), together with his student Robert Mills gave this demand for gauge invariance in the quantum mechanical setting a more general context and enlarged it into a requirement called *local gauge invariance.* This means that the rotation of the wave function at one point in space and time may be taken to be different from that at any other point in space and time. Yang and Mills demonstrated that if the quantum theory of particles is subjected to this general requirement, the theoretical existence of electric charge and of an electromagnetic field that obeys the Maxwell equations follows as a direct consequence. The entire existence of electricity and magnetism, and the fact that these fields satisfy the Maxwell equations, therefore, follow from the postulate that there has to be local gauge invariance.

The translation of this idea into a somewhat broader context, in which the allowed gauge transformations are based on larger groups than rotations in a plane, leads to the so-called *non-Abelian gauge field theories.* (Two rotations in a plane always commute, which means that it does not matter in what order they are performed. In three dimensions the order does matter, as we have already seen. This makes rotations in three dimensions a "non-Abelian" group. For the concept of a group and further details, see Chapter 10.) All the new field theories are of this kind. Their existence is generated by postulates of invariance under certain specified local gauge transformations.

The one force that has not yet been brought together with all the others into a unified field theory is the one that has been known the longest: the gravitational force. No one has yet succeeded in constructing a theory that combines gravitation with the quantum theory. At the present time there are, however, some ingenious attempts in that direction which have generated much mathematical activity but so far no experimentally accessible predictions. I am referring to *string theory*, which has received much attention from journalists. Apart from the fact that in this theory particles are not regarded as points, as in previous theories, but as extended strings, the fascinating idea here is that most of these theories are formulated in higher dimensions.

Physical reality, of course, exists in the four dimensions consisting of the one dimension of time and the three dimensions of physical space. In the 1920s there had been an imaginative but largely ignored attempt at unifying the electromagnetic field with gravitation by the Polish mathematician Theodor Kaluza and the Swedish physicist Oskar Klein. In order to do that, they formulated their theory in five dimensions, one time dimension and four space dimensions. The new theories were stimulated by the Kaluza–Klein theory, but most of them are formulated in many more dimensions, achieving a seemingly satisfactory combination of gravitation with all the other field theories and the quantum theory by working either in ten or in twenty-six space-time dimensions. The number of dimensions is reduced to the four of physical reality (three of space and one of time) by postulating that the extra dimensions curl up on themselves into very small circles.

In other words, our world is, in these theories, like a very thin worm hole, and of course we are unaware of the possibly large apple in which that worm hole is embedded. Our physical space consists of the three dimensions that extend on the wall along the length of the tunnel, and we are oblivious of the fact that every point in this space really consists of a tiny circle in higher dimensions. It is as though, in order to specify the position of a spider on the wall of the Lincoln Tunnel we would think it sufficient to tell how far it is from the entrance and neglected to say how high up on the wall it sits.

Needless to say, not all physicists are persuaded of the soundness of these ideas, and since the string theories are mathematically very difficult and have little chance of confrontation with experiments in the forseeable future, it may take a long time before they are either accepted or definitely rejected. In the meantime, however, a number of theoretical physicists and their brightest graduate students study this esoteric field. Some think that if one of these theories is experimentally confirmed, it will be *the theory of everything.* If, on the other hand, none is, the arcane knowledge thus gained may have purely mathematical significance, but physically it will be as useful as alchemy. (What sometimes happens, however, is that the mathematical tools developed for an unsuccessful theory later turn out to be useful in another context.)

Field theory, as we can see, has come a long way since the days of Faraday. Not only have our ideas of the forces changed during the last century and a half, but our ideas of what constitutes space and time have been profoundly altered. The vacuum, which by its very definition was *empty* nothingness, is now regarded as teeming with energy and virtual particles of many kinds. Perhaps we are close to inventing the theory of everything. But expressions of similar hubris by scientists in the past have always evoked tolerant smiles and scorn by succeeding generations. There is little reason to expect that our present arrogance will appear any less ludicrous to future scientists.

Let us return now to the oldest of the field theories, that of electromagnetism governed by the Maxwell equations. Of all the infinitely many solutions of these equations, perhaps the most important are those that describe light waves. There are, of course, many other kinds of waves we encounter in our daily lives, such as musical sound waves and water waves. What do all of them have in common and what properties distinguish them? These are the kinds of questions we want to discuss in the next chapter.

5

WAVES: STANDING, TRAVELING, AND SOLITARY

Waves of one kind or another can be found everywhere, and they have played a dominant role in physics for the last century and a half. In addition to sound and light waves, think of the waves of radio and television transmission, microwaves, water and seismic waves. Others are gravity waves, waves that transmit nerve impulses, and "particle waves" in quantum mechanics. Most of these have certain important features in common; others are distinguished particularly because they lack those features. This chapter will examine the ideas that form the basis of our description of the various kinds of waves. Some of the properties are recognizable from everyday experience; others are not so familiar but quite intriguing and even surprising. It is a testimony to the unifying power of mathematics that the same or closely related equations govern the behavior of quite disparate physical systems, producing analogous phenomena.

The Vibrating String

Let us begin with a string, made of gut, metal, or some other elastic material, clamped at both ends, as in a musical string instrument. The vibrations of such strings are called *standing waves;* they produce the sound waves in the air that we call music if we regard them as beautiful, and noise if ugly. The Greek philospher, mystic, and mathematician Pythagoras of Samos was the first to discover some 2,500 years ago that there was a definite relation between the lengths of harp strings and the musical notes and overtones they produce. But the first person to investigate these vibrations on the basis of Newtonian mechanics was the English mathematician Brook Taylor, in 1715. They were studied more generally forty years later by the Dutch-Swiss mathematical physicist Daniel Bernoulli. However, it was the English physicist John W. Strutt, better known as Lord Rayleigh, who made the largest contributions to the theory of sound and the theoretical study of the vibrations that give rise to sound waves. Lord Rayleigh was a rare instance of a hereditary peer who devoted himself to the pursuit of science, succeeding Maxwell, who was eleven years older than he, in the Cavendish professorship at Cambridge upon the latter's premature death. Let us take a look at the basic ideas.

In order to understand the equation that governs the vibrations of a flexible string, we apply the Newtonian equation of motion to each short piece of length L. Figure 22 shows a string that has the tension T, so that a small piece has the force $\vec{T}_1$ pulling on one end and the force $\vec{T}_2$ pulling on the other. However, since the string is not quite straight, these two forces do not point exactly in opposite directions and therefore do not quite cancel one another; the result is a small unopposed pull in the direction toward the horizontal straight line that is the equilibrium position of the string (that is, the position of the string at rest). If we assume that the displacement D of the string is small, then the resulting pull is vertically downward in the picture.

Now note that the two forces $\vec{T}_1$ and $\vec{T}_2$ are tangential to the curve of the string, which at the time t is described by the unknown function $D = f(t,x)$, so that they point in the directions of the slopes of f, which are given by the partial derivative f_x at the points x_1 and x_2, at the fixed time t, as we discussed in Chapter 4. As the resulting force on the short section of length L is the difference between the two tangential pulls $\vec{T}_1$ and $\vec{T}_2$, it will be equal to the tension T multiplied by L times the partial derivative of f_x. In other words, the force points downward in the vertical direction, and its magnitude is $-LT f_{xx}$. On the further assumptions that the string is quite flexible and that its weight is negligible, this is the only force on this small segment. If the mass of the string is M per unit length so that the mass of the piece of length L is LM, then the Newtonian equation of

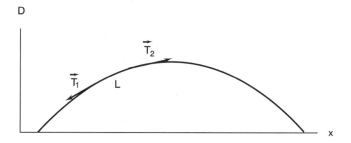

FIGURE 22. The tension-forces on a stretched string.

motion $\vec{F} = m\,\vec{a}$ for the short, almost horizontal section of string at the distance D from the x-axis becomes $LM\,D_{tt} = LT\,D_{xx}$. (Recall that the acceleration $\vec{a}$ is the second time-derivative D_{tt} of the distance D.) The length L of the string segment cancels out of this equation, and if we divide by M, writing as an abbreviation $c^2 = T/M$, we obtain the partial differential equation

$$D_{tt} = c^2\,D_{xx}$$

known as the *wave equation*. So long as it does not deviate very far from the horizontal straight line, the curve $D = f(t,x)$ that describes the shape of the vibrating string as a function of the time t and as a function of the distance x from one end must be a solution of this wave equation. If the string has a uniform density, which is usually the case, then c is a constant that can be calculated in terms of the tension and density of the string by the simple formula $c = \sqrt{T/M}$.

If the physical system is not one-dimensional like a string, but two-dimensional like the membrane of a drum, so that the displacement is a function of the two coordinates x and y in addition to the time, then the only change is that the partial derivative D_{xx} is replaced by $D_{xx} + D_{yy}$. If the system is three-dimensional, as for sound waves in the air or in water, or for electromagnetic waves such as light or radio waves, then three coordinates x, y, and z are needed, and D_{xx} is replaced by $D_{xx} + D_{yy} + D_{zz}$. Let us confine ourselves, to begin with, to the one-dimensional string.

The reasoning leading to the derivation of the wave equation for the vibrating string illustrates a characteristic feature of many arguments in theoretical physics: we did not attempt to describe the physical situation in its fullest precision with all details taken into account. Instead, in order to arrive at a simple description that captures the essence of the motion, we made certain approximations that left some aspects of the full physics out of the picture. Had we not assumed that the displacement of the string from its equilibrium is small, that the stiffness of the string and the force of gravity are negligible, the resulting equation would have been more realistic, but also very much more complicated and harder to analyze. Physicists always prefer a simple description of a phenomenon to a full and exhaustive one, even at the price of having to make later corrections to take other perturbing influences into account before comparison of the theory with experiments and observation can be made. This act of simplification and abstraction is a good example of a scientist's imagination coming into play, and it is one of the reasons why the mathematical description of natural phenomena is not a simple matter of "telling it like it is."

But let us return to our vibrating string of a given length l, held fast at both ends. The curve that describes its displacement from the horizontal line, $D = f(t,x)$, must satisfy the wave equation, which is a partial differential equation of the second order. The solution of this equation requires that we supplement it with boundary conditions or initial conditions, or both. The fact that the string is clamped at both ends provides us with such conditions: we must require that at all times the displacement is equal to zero at $x = 0$ and at $x = l$; so the appropriate boundary conditions are $f(t,0) = f(t,l) = 0$.

In order to find the solutions of this mathematical problem, we have to make a small detour that will lead us to functions that arise in many other contexts of physics.

Sine and Cosine Functions and Their Derivatives

Figure 23 shows a circle of radius 1 about the coordinate origin. Consider the x and y-coordinates of the point P as a function of the angle a, measuring a in radians rather than in degrees. This simply means that a is measured by the distance from the point Q to the point P along the circle. Since the circumference of the circle is 2π (its radius is 1), 360° corresponds to 2π radians, 90° corresponds to $\pi/2$, and so on. The functional dependence of x upon a defines a function called *cosine,* written $x = \cos a$, and the dependence of y upon a is called *sine*, written $y = \sin a$. As you can see from the figure, as a increases from zero, so does $y = \sin a$, while $x = \cos a$ decreases from its maximal value 1; when a reaches $\pi/2$, that is, 90°, $\sin a$ has increased to its maximal value of 1, and $\cos a$ has decreased to 0. As a further increases from $\pi/2$ to π, $\cos a$ decreases to −1 and $\sin a$ to 0.

WAVES: TRAVELING AND SOLITARY

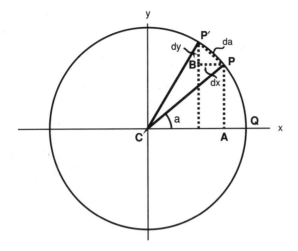

FIGURE 23. Definitions of the sine and cosine functions, and
of their derivatives. (See text for an explanation.)

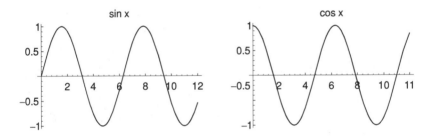

FIGURE 24. Plots of the sine and cosine functions.

The resulting curves are plotted in Figure 24. Since the entire process
along the circle repeats itself after one revolution, both curves are
periodic with the period 2π.

Consider some further details in Figure 23: in the picture, da is
meant to be a small increase in the angle a, and dx and dy are the
corresponding small differences in the x and y coordinates as the
point P shifts to P'. A bit of elementary geometry tells you that since,
for a small increment da, the straight line from P to P' is tangent to
the circle, the line PP' is perpendicular to the radial line CP and the
triangle CAP is similar to the triangle PBP'. From this we may conclude that the ratios of corresponding side lengths are equal: since for
small da we have $da = PP'$, we may write these ratios as

$dx/da = -y/1$ (note that dx is negative!) and $dy/da = x/1$. Recall now from Chapter 2 that for small da the ratios dx/da and dy/da are the derivatives of the two functions $x = \cos a$ and $y = \sin a$. Therefore we have found that $(\cos a)_a = -\sin a$ and $(\sin a)_a = \cos a$. What is more, inserting one of these equations into the other, we find that $(\cos a)_{aa} = -\cos a$ and $(\sin a)_{aa} = -\sin a$. In other words, both of these functions solve the second-order differential equation $f'' = -f$. For a simple extension of this result we use the fact that if the independent variable is ka instead of a, where k is some constant, then $f'(ka) = [f(ka_2) - f(ka_1)]/(ka_2 - ka_1) = (1/k) [f(ka_2) - f(ka_1)]/(a_2 - a_1) = (1/k)f_a(ka)$ and hence $f_a(ka) = kf'(ka)$; therefore $[\cos(ka)]_a = -k\sin(ka)$ and $[\sin(ka)]_a = k\cos(ka)$. This leads us to conclude that the functions $f = \cos(ka)$ and $f = \sin(ka)$ solve the differential equation $f_{aa} = -k^2 f$ if the independent variable is a.

Solving the Wave Equation for the String

It is now time to return to the wave equation $D_{tt} = c^2 D_{xx}$ with the boundary condition $D(t,0) = D(t,l) = 0$. This boundary condition, which is supposed to hold for all t, is easy to meet if we try a function that is simply a product of a function of t times a function of x, $D(t,x) = F(t)f(x)$, with f chosen so that $f(0) = f(l) = 0$. Such a product, in fact, works, provided that f solves the differential equation $f'' = -k^2 f$ and F solves the equation $F_{tt} = -(kc)^2 F$. We have already found a solution of the equation $f'' = -k^2 f$ that meets the boundary condition $f(0) = 0$, namely $f(x) = \sin(kx)$. Therefore, all we have to do in order to meet the other boundary condition $f(l) = 0$ is to choose k in such a way that $\sin(kl) = 0$. But we also know that $\sin\pi = 0$, and since the function is periodic, $\sin 2\pi = \sin 3\pi = \ldots = 0$. So if we give k any of the values $\pi/l, 2\pi/l, 3\pi/l, \ldots$, both boundary conditions are satisfied and the problem is solved. The function F, which has to solve the equation $F_{tt} = -(kc)^2 F$, can now be chosen to be $F(t) = \sin(kct)$ or $F(t) = \cos(kct)$, or any combination of the two, depending on the initial condition. For instance, if we require that at the initial time $t = 0$ the string should be in its equilibrium position $D = 0$, we must take $F(t) = \sin(kct)$.

Our principal conclusion is that the solution of the wave equation, with the given boundary and initial conditions, is $D(t,x) = \sin(kct)$ $\sin(kx)$, and k must be taken to have one of the values $\pi/l, 2\pi/l, 3\pi/l, \ldots$ Because the function $\sin(kct)$ is a periodic function of t with the period $2\pi/kc$, this solution is a periodic function of the time whose frequency (which is the number of vibrations exe-cuted per second and therefore equal to the reciprocal of the period of the vibration) depends on our choice of k. If k has its smallest possible value π/l, the frequency has the smallest value $c/2l$.

There is an important lesson contained in our result. We know that the second-order differential equation $f'' = -k^2 f$ requires two bound-ary conditions for its solution; in Chapter 2 we took these to be $f(0)$ and $f'(0)$. Here, however, we assigned the so-called "homoge-neous" boundary conditions $f(0) = f(l) = 0$, and we found that un-less k^2 has certain specific values, which are called the *spectrum*, the only solution is $f(x) = 0$ (which mathematicians call the "trivial solu-tion"). This is an instance of a much more general mathematical fact: certain differential equations with "homogeneous boundary condi-tions" have "nontrivial" solutions (that is, solutions that don't simply vanish) only for special values of a parameter that appears in them. Such differential equations appear in many other physical contexts, like the propagation of microwaves in a wave guide, in resonating cavities, and above all in many areas of quantum mechanics, to which we shall return. We may consider it a matter of some philosophical significance that even though we began with an assumption that the material studied, the vibrating string, forms a continuum and that time varies continuously (so that we could set up a differential equa-tion for the motion), we were led to *discrete* values for the kinds of motion that are possible. The appearance of a *spectrum* of allowed values has a profound effect on many physical processes, one exam-ple of which is the vibrating string we are now discussing.

The two simplest vibrations of the string are shown in Figure 25 The first is such that half a wave fits into the length l, so that the wavelength λ is $2l$; the next has a full wave in l and the wavelength is l. In other words, the wavelength λ is related to k by $\lambda = 2\pi/k$, and the permitted wavelengths are $2l, l, \frac{2}{3}l, \frac{1}{2}l, \frac{2}{5}l, \dots$ The frequency ν of the vibration that goes with a given value of k, as we saw above, is $\nu = kc/2\pi$; therefore $\nu = c/\lambda$, and the frequencies produced by the string are $c/2l, c/l, 3c/2l, 2c/l, \dots$

The Sound of Music

An oscillation of the string is directly transmitted to the surrounding air and causes a vibration there with the same frequency; this we hear as *sound*, and a different frequency is heard as a different pitch. Thus, the fundamental note produced by a string of length l, density M, and

FIGURE 25. Snapshot of the fundamental variation
of a stretched string and its first harmonic.

tension T corresponds to the frequency $c/2l$, where $c = \sqrt{T/M}$. The higher the tension of the string, the higher is the pitch of the resulting note; the heavier the string, the lower the note. The same string also produces a note that is an *octave* higher, whose frequency has the ratio 2:1 to the lowest note produced, a *fifth,* whose frequency has the ratio 3:2, a *fourth,* with the ratio 4:3, and so on. (That such ratios of frequencies are responsible for the production of harmonics was established in the seventeenth century by the Minimite friar Marin Mersenne, a friend of Descartes and Fermat and a correspondent with many other mathematicians of his time.) We have found, therefore, that the solution of the mathematical problem posed by the wave equation together with homogeneous boundary conditions completely explains the production of musical notes and their harmonics by the vibrations of a string that is fixed at both ends.

Here you see the reach of the power of mathematical concepts: the most advanced and abstract application of the fact that a vibrating string swings with certain characteristic frequencies occurs in the string theories of contemporary quantum field theory, in which the most fundamental particles in nature are thought of as tiny strings in a higher-dimensional space (see Chapter 4). By the rules of the quantum theory, the spectrum of these vibrations is meant to account for the masses of the elementary particles. At this point, however, there is no experimental evidence to support these very imaginative ideas, and we return to the more ordinary, though no less important, subject of music.

A string on a musical instrument produces not only a single pure note but also some overtones at the same time. That is because the string can produce not only the fundamental note, or the octave above, or the fifth, and so on, but in fact all the harmonics simultaneously. Furthermore, it can produce them at any strength desired (within limits). The reason for this is that the wave equation is both *homogeneous* and *linear.* The first of these terms means that if $D(t,x)$ is a solution, then so is $aD(t,x)$ for any arbitrary constant a. This is so because it follows from the definition of the derivative that $(aD)_{xx} = aD_{xx}$, and similarly for the time-derivative. To say that the

WAVES: TRAVELING AND SOLITARY

equation is linear is to say that it contains the dependent variable D only as a first power and not squared, or in higher powers; $4 + 7D$ is a linear function of D, but $6 + 9D + D^2$ is not. This has the consequence that if the waves $f(t,x)$ and $g(t,x)$ both separately solve the wave equation, then so does their sum $f + g$, which is also called the *superposition* of the two; we say that the wave equation obeys the *principle of superposition*. Since the boundary conditions are also homogeneous and linear, the string can produce all the harmonics of the fundamental note simultaneously. The strength with which it produces them, of course, depends on the initial condition, that is, how it is plucked or struck.

As is almost always true with mathematical ideas, the concept of superposition, which works for the solutions of linear equations, can be fruitfully generalized. Let us superpose an *unlimited number* of different solutions of the wave equation with different allowed values of k, with arbitrary strengths, and with the function $F(t)$ chosen to be $cos(kct)$,

$$D(t,x) = a \cos\left(\frac{c\pi t}{l}\right)\sin\left(\frac{\pi x}{l}\right) + b\cos\left(\frac{2c\pi t}{l}\right)\sin\left(\frac{2\pi x}{l}\right) + \ldots$$

At the initial time $t = 0$ the displacement is then given by

$$D(0,x) = a\sin\left(\frac{\pi x}{l}\right) + b\sin\left(\frac{2\pi x}{l}\right) + \ldots,$$

because $\cos 0 = 1$. Jean Baptiste Fourier—who, though much embroiled in the French Revolution, still had the time and energy to do important work in mathematics—found that such a sum, now called a Fourier series, can, by proper choice of the coefficients a, b, ... be made to add up to any arbitrary given function. What is more, if the function $D(0,x)$ is given, the coefficients a, b, ... , which are called the *amplitudes* of the various sine waves, can be easily calculated. This means that the strength with which any harmonic is excited is completely determined by the initial configuration of the string, that is, by the way it is plucked. Similarly if the string is struck rather than plucked: in that case its initial configuration is the equilibrium position $D(0,x) = 0$ and the initial velocity of every point is given.

Traveling Waves

In addition to standing waves, there are other kinds of string oscillations. For example, if you tie one end of a long rope to a tree, hold the other end in your hand, and begin to shake it, you see a wave moving along the length of the rope. This phenomenon is described by a *traveling-wave* of the wave equation. Such solutions are easily found. It is not hard to demonstrate that if we take any arbitrary function $f(s)$ and let the independent variable s be given by $s = x - ct$, the function $D(t,x) = f(x - ct)$ satisfies the wave equation. If, as t increases, x is also made to increase in such a way that $s = x - ct$ stays fixed, then D stays fixed. In other words, as time goes on, the plot of the function D keeps on shifting rigidly to the right and the "wave" described by $f(s)$ travels to the right without changing its shape. In order to keep s fixed, x has to advance with the speed c; therefore the speed with which the wave travels is given by c. At the same time, because the equation is homogeneous, the amplitude or strength of the wave is quite arbitrary.

Of course, if the wave, as a function of the time, is to be sinusoidal (that is, in the shape of the sine function), as it will be if it is excited by a single harmonic of frequency v of a clamped string, then its spatial shape has to be sinusoidal too, like $\sin[2\pi v(x - ct)/c]$. (The constant c that appears here need not have the same value as the constant c in the wave equation for the clamped string which may have excited this traveling wave.) We could also just as well replace $s = x - ct$ everywhere by $s = x + ct$; in that case the wave would be traveling to the left rather than the right, because as t increases, x has to decrease in order for $s = x + ct$ to remain fixed.

The salient characteristics of the traveling-wave solutions of the wave equation, in sum, are that they have arbitrary amplitudes (and shapes) and that their speed is fixed by the number c that appears in the equation. Moreover, since the wave equation is homogeneous and linear, they are subject to the *superposition principle:* any two of them can be added and the sum will again be a solution. Because of Fourier's theorem, we may, in fact, always think of any solution of the wave equation as a sum of sinusoidal waves, each having a fixed frequency v and wavelength $\lambda = c/v$.

Interference

A number of special phenomena are produced by the superposition of traveling waves. Let us, for example, superpose two waves that travel in the same direction and have the time dependence $\cos(vt)$. Though they must have the same wavelength $\lambda = c/v$, however, their spatial dependence on the distance x may differ by a *phase*. By this we mean that one of them is of the form $a \sin(kx)$, whereas the other is of the form $b \sin(kx - p)$, where p is some constant; the maxima of the second are displaced by p/k from those of the first. The result of the sum of the two will be a wave that has the same wavelength $\lambda = 2\pi/k$ but that exhibits *interference*. Two special cases are shown in Figures 26 and 27. In Figure 26 the phase difference p is equal to zero. The result is a new wave whose amplitude is the sum of the amplitudes of the two waves: $a\sin(kx) + b\sin(kx) = (a + b) \sin(kx)$; this is called *constructive interference*. Figure 27, on the other hand, shows a case in which the phase difference is $p = \pi$. It is clear from the picture that the result is a new wave whose amplitude is the difference between the two original amplitudes: $a\sin(kx) + b\sin(kx + \pi) = (a - b)\sin(kx)$. This is called *destructive interference*. If the two amplitudes a and b happen to be the same, then the superposition of the two waves is simply zero; the wave has totally disappeared. If the phase difference p is somewhere between 0 and π, the result is intermediate between these two extremes.

There is another interference phenomenon of a different kind. Consider the time dependence of two waves of slightly different frequencies arriving at the same location. Their superposition will then be of the form $a\sin(2\pi v_1 t) + b\sin(2\pi v_2 t)$, and such a sum, in which v_1 and v_2 differ by a small amount, is plotted in Figure 28. (In these illustrations we have chosen $a = b = 1$.) Notice that there is a slow variation of the amplitude of the sum; this variation, called *beats*, is what you hear when two tuning forks are vibrating at the same time and one of them is slightly out of tune.

The occurrence of interference is a significant characteristic of waves and has important applications. However, before looking at some of these applications, we first have to realize that the wave equation appears in many different physical contexts, and its solutions in two and three dimensions have characteristics that are quite analogous to those in one dimension.

Vibrating Membranes

The two-dimensional analogue of the vibrating string is a vibrating membrane, as on a drum or a metallic plate. The equation satisfied by the displacement of the membrane from equilibrium is the two-dimensional wave equation and the boundary condition is that the displacement should vanish on the boundary, whose shape must be given. The French mathematician Siméon Denis Poisson was the first

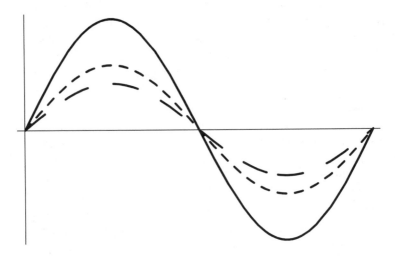

FIGURE 26. Constructive interference; the solid curve
is the sum of the two dotted curves.

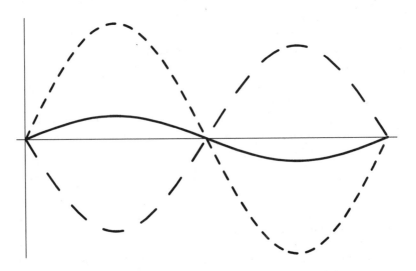

FIGURE 27. Destructive interference; again the solid curve
is the sum of the other two.

WAVES: TRAVELING AND SOLITARY

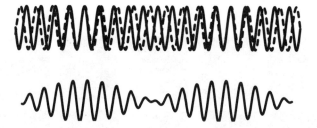

FIGURE 28. The upper curves are two sinusoidal waves of slightly different wavelengths; the lower curve shows the beats produced by the sum of the waves in the upper figure.

to describe these vibrations successfully in 1829, and their general theory was given by Rudolf Clebsch thirty-three years later. The result is again a spectrum of allowed frequencies. The possible spatial configurations are now more complicated, with the *nodes,* at which the vibrating surface is always at rest in its equilibrium position, forming figures that can be made visible by sprinkling sand on the surface. The sand remains in place only where the membrane is at rest and is thrown off everywhere else.

This method of making nodal figures visible was invented in the late eighteenth century by Ernst Chladni, a German lawyer who loved the science of acoustics. Clamping a metal plate in the center, he sprinkled fine sand on it and made it vibrate by stroking its edge with the bow of a violin. The results were a great variety of beautiful web-like patterns, which so impressed Napoleon that he established a prize for the first person who could explain them mathematically. (Some examples are shown in Figure 29.) With Napoleon in exile, the prize was awarded in 1816 to Sophie Germain, who, as a woman, had not been allowed to attend a university. However, it took many years of further efforts to get the theory quite right. The shape of the nodal figures depends, of course, on the shape of the plate or membrane, becoming quite chaotic if that shape lacks symmetry.

As in the case of the vibrating string, the oscillations of the plate or drum head are communicated to the surrounding air and produce sound waves whose pitch is given by the frequency or frequencies produced by the drum. In this case the allowed spectrum depends not only on the weight, tension, and size of the membrane but also on its shape. If that shape is more complicated than a circle or a rectangle, the spectrum can no longer be given by a simple formula, as it can for the uniform string, but has to be computed numerically. (The

FIGURE 29. Some Chladni figures: the white curves are the nodes of vibrations of a clamped plate.

WAVES: TRAVELING AND SOLITARY

relation between the shape of the boundary of a vibrating membrane and its spectrum gives rise to an interesting "inverse problem" that has been solved by mathematicians; Mark Kac described both the problem and its solution in a paper entitled "Can one hear the shape of a drum?")

Vibrations in Three Dimensions: Sound and Light

In three dimensions, the wave equation governs many phenomena of various kinds. There are the sound waves, which Lord Rayleigh studied most thoroughly in 1877 in his treatise *The Theory of Sound.* In air and liquids, these are compression waves that are produced by *longitudinal* vibrations of molecules around their equlibrium positions, first described as such by Isaac Newton. Just as a series of traffic obstructions on a highway will produce waves of larger and smaller densities of cars, so such vibrations give rise to waves of compressions and decompressions, which our ear registers as sound; the higher the frequency, the higher the pitch we hear. To Newton's embarrassment, however, the formula he had found for the speed of sound propagation in air consistently disagreed with measured results, and he tried to repair the damage in the second edition of the *Principia* by unconvincing ad hoc modifications. In truth, he had made an unjustified initial assumption that was not corrected until a hundred years later by Laplace.

The range of frequencies that are audible to the human ear runs roughly from 20 Hz to 20,000 Hz, a range of about ten octaves. (Hz stands for the unit *hertz,* named after the German physicist Heinrich Hertz, which is one oscillation per second.) If the deviations of the air pressure from its ambient equilibrium are relatively small, then, like small vibrations of a string, the sound waves satisfy the wave equation. Consequently the propagation speed of a sound wave does not depend on its frequency (otherwise different notes produced by an instrument would take different times to arrive at our ear, and a coherent reception of music would be impossible), and it is not affected by the velocity at which the source of the sound is moving. In the open air, the velocity of sound, which was first measured accurately in 1738 by members of the French Academy, is approximately 1,000 feet per second; liquids and solids propagate sound with a higher velocity. In a vacuum, of course, there can be no sound, since compression waves require a medium to be compressed.

The phenomenon of destructive and constructive interference of sound waves in a concert hall is familiar to most of us. "Dead spots" and distortions can arise from the interferences between waves that are reflected from various parts of the architectural structure at different distances from the listener, producing different phases. (If a wave that arrives at the listener's ear directly from an instrument at a distance D is proportional to $\sin(2\pi v t) \sin(kD)$, a wave that has been reflected and has traveled a distance D' is proportional to $\sin(2\pi v t) \sin(kD') = \sin(2\pi v t) \sin(kD + p)$, where the phase difference is given by $p = k(D' - D) = 2\pi(D' - D)/\lambda$; the superposition of the two waves leads to interference. Note that p depends on λ; therefore different tones suffer different interferences and the resulting sound is distorted.) The pulsating and irritating phenomenon of beats, which can arise when two instruments, one of which is slightly out of tune, are played simultaneously, is another example of interference in sound waves.

Consider next the theory of electromagnetism. Maxwell's equations for the electromagnetic field, developed a few years before Rayleigh's treatise, also have solutions that at the same time solve the wave equation. In this case, however, the oscillating electric and magnetic fields are perpendicular to the direction of propagation of the wave (as well as perpendicular to one another); therefore, these waves, examples of which are light waves, radio waves, microwaves, x-rays, and heat rays, are called *transverse*. The only difference between these various waves is that they have different wavelengths; in the case of visible light, the wavelengths range between 4×10^{-5} cm and 7×10^{-5} cm. The light of different wavelengths is seen in different colors, violet and blue being the shortest and red and orange the longest. The wavelengths of x-rays are very much shorter, those of infrared heat rays longer, and those used for radio and TV broadcasts very much longer still; long radio waves may have wavelengths of hundreds of miles, while x-ray wavelengths go down to about 10^{-9} cm. If a ray of light is such that its oscillating electric field lies in a plane, it is called plane *polarized;* a polaroid filter, the kind used in photography and sunglasses, produces such polarized light by passing only waves whose polarization is in a given direction. Sound waves, being longitudinal, have no analogous property.

Because the Maxwell equations in free space are linear and homogeneous, electromagnetic waves obey the superposition principle, which implies that, apart from interference, two light waves do not scatter or disturb each other. It is only when the quantum theory is

WAVES: TRAVELING AND SOLITARY

taken into account in the manner described in Chapter 4, when the resulting equations of quantum electrodynamics are nonlinear, that there is such a thing as scattering of light by light. This effect is called Delbrück scattering; since it is extremely small, it is hard to observe and its existence has not as yet been experimentally verified. (Max Delbrück, after whom this effect is named, was the first to calculate its magnitude; he was at that time a physicist but later switched his main field of interest to microbiology, where he and his "phage group" at Caltech became extremely influential.)

Destructive and constructive interference between light waves is not hard to observe. We see this phenomenon, for example, in the colored bands visible in oil slicks on a water surface, which are caused by interference between the light reflected from the two surfaces of the thin layer of oil. They are colored because the resulting phase difference depends on the wavelength. If the monochromatic light emerging from a pinhole is made to pass through two thin parallel slits not very far apart, the image seen on a screen consists of a succession of narrow dark and light bands called *fringes* caused by constructive and destructive interference of the light waves that pass through one of the slits with those that pass through the other. If the light is not monochromatic but white (which is a mixture of all different colors), the fringes are colored because the position of the interference maxima depends on the wavelength, so that different colors have their maximal intensities at different places. The English physicist Thomas Young, at the beginning of the nineteenth century, was the first to perform this unequivocal demonstration that light consists of waves. (This was still a controversial thought; Newton had pictured light as consisting of small particles.) If we replace the screen having two slits with one having a series of many narrowly-spaced fine parallel slits—a device called a *diffraction grating*—the image seen on a projection screen is again a series of interference fringes whose spacing depends on the color of the light. This effect is, in fact, a practical method for measuring the wavelength of the incident light.

The Michelson–Morley experiment, which played an extremely important role in the modern history of physics because it was the experimental starting point for the theory of relativity, was already mentioned in Chapter 4 and will be described in some detail in Chapter 6. It was designed to measure our velocity with respect to the "luminifferous ether" by measuring the local speed of light in various directions. I mention it here because the precision needed for its success depended entirely on the observation of fringes caused by the interference of two light rays of different phases.

Perhaps the phenomenon that best unites the wave nature of sound with that of light is known as the Doppler effect. Who has not felt a sense of nostalgia caused by the sound of a train whistling in the night? If you have ever heard it close by, you will recognize a characteristic fact about it: its pitch drops abruptly as the engine passes. That is the *Doppler effect,* named after the Austrian physicist Johann Christian Doppler. Its explanation rests entirely on the wave nature of sound. If the advancing sound wave is caused by the periodic "kicks" of a vibrating source moving with the speed v, the period T of the vibration must be equal to the time it takes from one crest of the resulting wave to the next. Therefore, the distance between the crests must be $\lambda' = (c - v)T$ if the sound and the whistle move in the same direction, and $\lambda'' = (c + v)T$ if in the opposite direction. This implies that the wavelength of the sound emitted forward is shorter than that emitted backward. The frequency of the wave received by the ear of a listener at rest in the air, of course, is given by $v' = c/\lambda'$ or $v'' = c/\lambda''$, depending on whether we are situated in front or in back of the moving source. As the train approaches us, we hear the sound emitted in the forward direction (the wavelength of this sound is shortened; its frequency is therefore greater and its pitch higher); as the train recedes, we hear the sound emitted in the backward direction (the wavelength is stretched and of that sound its pitch is therefore lowered). Consequently we hear an abrupt change in pitch when the locomotive passes.

This phenomenon has an exact analogue for light waves, and astronomers make use of it in the interpretation of the observed *red shift* of the light coming from stars in distant galaxies. The light emitted by excited atoms has a frequency spectrum that is a recognizable characteristic for each element, and the light that reaches us from the stars contains the same frequency patterns that we would observe from atoms on the earth. However, in most cases these patterns are shifted toward the red end of the spectrum, that is, toward the longer wavelengths, and each star has its own characteristic red shift. It is customary to interpret that shift as a Doppler shift that is analogous to the lowering of the pitch of a receding train whistle. According to that interpretation, the red shift is indicative of a star's speed of recession; in fact, the amount of the shift is proportional to the recessional velocity. Though some mavericks are skeptical of this interpretation of the red shift, most astronomers accept it and conclude from measurements of red shifts that the universe is expanding, and that all parts of it are receding from one another; the further away an object is, the faster it moves. A large part of our presently accepted cosmology depends, therefore, on the acceptance of the Doppler effect for light waves as the explanation of the observed red shift, which at the same time provides another example of the unifying force of mathematical thinking.

Matter Waves

By the end of the nineteenth century, two major areas of physics, the field of electricity and magnetism and the field of acoustics, as described above, were largely concerned with linear wave phenomena governed by the wave equation. Acoustics was, of course, a part of Newtonian mechanics and its description of small vibrations. The remainder of Newtonian mechanics, however, does not deal with linear equations, whose solutions satisfy the superposition principle, so that a large part of physics, still very much at the forefront of research at the beginning of this century, was "nonlinear." This situation changed dramatically when Louis De Broglie suggested in his doctoral thesis in 1924 that, just as, according to Einstein's quantum theory, light had a particle-aspect in addition to its classical wave nature, so particles such as electrons should also have a wave-aspect. The wavelength of a particle of mass m and velocity v, he proposed, should be $\lambda = h/mv$, where h is Planck's constant, the same constant that, according to Einstein, entered into the relation $E = h\nu$ between the frequency of light and the energy of a photon; this wavelength is now called the particle's *de Broglie wavelength*. Two American physicists, Clinton Davisson and Lester Germer, experimentally confirmed this astonishing hypothesis three years later by employing a diffraction grating in the same manner by which light had been demonstrated more than a century earlier to be a wave phenomenon, though the fact that the electrons they used had a de Broglie wavelength very much shorter than that of light made the experiment more difficult.

The wave nature of matter became an integral part of Schrödinger's version of quantum mechanics, which for many years was also called wave mechanics, and it follows as a consequence of the Schrödinger equation, the fundamental equation that incorporates the quantum phenomena. Even though we are, in this case, not describing anything that physically vibrates, and this differential equation is not quite the

119

same as the wave equation, the Schrödinger equation too is linear and homogeneous, so that its solutions satisfy the superposition principle. The resulting interference phenomena, such as interferences between electrons passing through two parallel slits as in Young's experiment for light, are part and parcel of quantum-theoretical arguments and explanations, making up the principal difference between classical mechanics and quantum mechanics. They also give rise to the puzzling *wave–particle duality* that makes the quantum theory intuitively difficult.

In our discussion of vibrating strings and membranes, we found that the wave equation with homogeneous boundary conditions leads to a discrete spectrum of allowed wavelengths; the same holds for the Schrödinger equation. Thus the electrons in an atom, for example, whose waves satisfy the Schrödinger equation, can exist only with a specific spectrum of wavelengths and hence of velocities or energies. These are the discrete energy levels of the atom which Niels Bohr first postulated without proper justification. When an electron in an atom descends from an upper to a lower level, it gives birth to a photon whose energy E is equal to the difference between the two energy levels of the parent atom, and the frequency v of the thus emitted light is related to that energy by Planck's constant h and Einstein's equation $v = E/h$. In this way, the Schrödinger equation completely explains the characteristic colors of the light emitted by atoms of different chemical elements. These identifying spectral lines of the light emitted by atoms on distant stars, when passed though a diffraction grating and projected on a photographic plate in a laboratory on earth, are the "fingerprints" that are observed to be red-shifted, and thus give evidence both that the stars contain the same elements as the earth and that they are receding from us at great speeds. Most of the "discreteness" that characterizes the quantum theory as distinct from its classical predecessors rests on the discreteness of the spectrum of the Schrödinger equation with homogeneous boundary conditions.

From the beginning of the twentieth century on, research in particle physics has used almost exclusively the methods and ideas of the quantum theory. It can therefore be fairly said that for almost seventy-five years physics was dominated by *linear* mathematical methods, which on the whole are much simpler than nonlinear ones. The dominance of linear equations in physics, however, has been increasingly challenged during the last twenty five years. One of the challenges originates from the growing study of chaos in dynamical systems, as outlined in Chapter 2. The other has come from investigations of a wave phenomenon that was first observed early in the nineteenth century but whose full investigation had to wait for the introduction of the electronic computer.

Solitons

The Scottish naval architect John Scott Russell described the "most beautiful and extraordinary phenomenon" he observed on that "happiest day of [his] life" in 1834 in the following memorable way:

I was observing the motion of a boat which was rapidly drawn along a narrow channel by a pair of horses when the boat suddenly stopped—not so the mass of water in the channel which it had put in motion: it accumulated round the prow of the vessel in a state of violent agitation, then suddenly leaving it behind, rolled forward with great velocity, assuming the form of a large solitary elevation, a rounded smooth and well defined heap of water, which continued its course along the channel apparently without change of form or diminution of speed. I followed it on horseback, and overtook it still rolling on at a rate of some eight or nine miles an hour, preserving its original figure some thirty feet long and a foot to a foot and a half in height. Its height gradually diminished and after a chase of one or two miles I lost it in the windings of the channel. Such in the month of August 1834, was my first chance interview with this singular and beautiful phenomenon which I have called the Wave of Translation.

These solitary water waves were observed to behave quite differently from waves described by the wave equation. First of all, they could not have arbitrary shapes; furthermore, and very importantly, though there could be large and small "great solitary waves," as he later called them, their speed depended on their size. "If such a heap," to quote Russell again,

be by any means forced into existence, it will rapidly fall to pieces and become disintegrated and resolved into a series of different waves, which do not move forward in company with each other, but move on separately, each with a velocity of its own, and each of course continuing to depart from the other. Thus a large compound heap or wave becomes resolved into the principal and residuary waves by a species of spontaneous analysis.

121

These were observed facts, and there was no known equation to describe and explain them.

Having ascertained that no one had succeeded in predicting the phenomenon which I have ventured to call the wave of translation . . . it was not to be supposed that after its existence had been discovered and its phenomena determined, endeavours would not be made . . . to show how it ought to have been predicted from the known general equations of fluid motion. In other words, it now remained for the mathematician to predict the discovery after it had happened, i.e. to give an *à priori* demonstration *à posteriori*.

As Russell laments, "We accordingly find that a theory of the wave of [translation] . . . is still wanting, a worthy object for the enterprise of a future wave-mathematician."

It took about fifty years for such wave-mathematicians to rise to the challenge: two Dutch mathematicians, D. J. Korteweg and G. de Vries, found a partial differential equation that does justice to the solitary waves; this is now generally referred to as the *KdV equation*. Based on the fundamental laws of fluid dynamics, it was designed to describe water waves in a shallow one-dimensional channel, but it has since been found to govern many other physical phenomena. Certain kinds of hydromagnetic waves, acoustic waves in some crystals, plasma waves, and pressure waves in liquid-gas bubble mixtures are all examples that are described by solutions of the KdV equation.

The characteristics of the Korteweg–de Vries equation are quite different from those of the wave equation. Most important, in addition to linear terms, it contains the product of the dependent variable and its spatial derivative; it is therefore both inhomogeneous and nonlinear, and its solutions do not satisfy the superposition principle. We cannot add two solutions to get a new solution, and we cannot multiply solutions by arbitrary constants to get another solution. Furthermore, if we search for solutions that move along without changing their shape, such as Russell's great solitary wave, we find that such solutions do exist, but their shapes are not arbitrary. The shape of the solitary-wave solution of the KdV equation is shown in Figure 30. It can be of arbitrary height, but as we change its height, its thickness also changes correspondingly. What is more, a wave of a given size has a velocity that is fixed; large waves move faster than small ones, just as Russell observed in his Dutch canal. These characteristics of the solitary-wave solutions of the KdV equation are the result of the fact that the equation, in contrast to the conventional

WAVES: TRAVELING AND SOLITARY

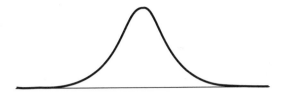

FIGURE 30. A solidary-wave solution of the KdV equation.

wave equation, is nonlinear. They describe exactly the properties of the waves that Russell observed in nature, as they were intended to do.

This is where matters stood for about seventy years, from the invention of the KdV equation in 1895 until 1965. The equation being nonlinear, and the theory of nonlinear differential equations being much less well developed mathematicically than that of linear equations, its general solution was not known. However, in the mid-1960s, a group of applied mathematicians, under the leadership of Martin Kruskal at Princeton University, put the KdV equation on a computer and did "computer experiments." Among other things, they played with the following simple idea: Suppose we start with two solitary waves, one large and the other small, far apart from one another (see Figure 31). Since they are far apart, each will be almost zero where the other one has its maximum. Therefore the product of the solution times its derivative will become essentially equal to the sum of two such products, one for each of the two waves, and the product of one times the other will almost vanish. As a result, the sum of two such solitary wave solutions whose centers are far apart will almost exactly solve the KdV equation, and as the two waves move along, their shapes will remain unchanged. However, one being large and the other one small, the larger one will move faster; if it was initially behind, it will catch up, and the centers of the two waves will approach one another. As the waves get closer, their sum will no longer solve the nonlinear KdV equation and the equation will force them to change their shapes; as they merge with one another, their shapes may change drastically. The nonlinearity makes the two waves *interact* with one another. This is what we would qualitatively expect, and this is what the computer confirms.

The interesting question now is: what happens to these waves as time goes on? Will they simply wash each other out, will they oscillate wildly and disappear? The answer was not known, and the computer experiment showed that what happens is quite astonishing. The

WAVES: TRAVELING AND SOLITARY

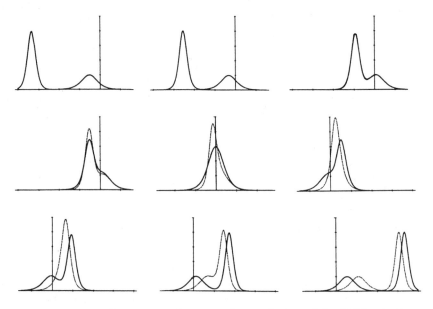

FIGURE 31. The motion of a two-soliton solution of the KdV equation; the dotted curves are superpositions of the two corresponding solitary waves as they would move without the other one present.

two waves that started out as solitary waves far apart "do their thing" as they get close to each other and may change shape. But after a while they begin to separate again. As time goes on, they emerge with exactly the same shapes and velocities they had initially, the bigger one now in front, as in Figure 31. The only remnant of the fact that they interacted at one time is a shift in their centers; the bigger of the two is shifted forward, compared with where it would be at a given time if the other one were not there, and the smaller one is shifted backward.

The computer showed this remarkable phenomenon every time two or more solitary waves collided. If any number of solitary waves traveling in the same direction approach one another, initially far apart, the same number of waves emerges, with unchanged shapes and velocities, their centers shifted and in the order of their velocities, with the fastest one out in front. These waves, which appear to have a permanent identity they retain even after collisions in which they temporarily merge as an oscillating, pulsating clump, were named *solitons*. Subsequent to their original computer-based discovery, Kruskal and his co-workers were even able to offer an analytic explanation, which was based on relating the (nonlinear) KdV equation to

WAVES: TRAVELING AND SOLITARY

the (linear) Schrödinger equation. Since these waves had nothing whatever to do with the quantum theory, their connection with the Schrödinger equation was every bit as surprising as the soliton phenomenon itself. The same Schrödinger equation with homogeneous boundary conditions, whose spectrum explains the energies of atoms and the color of the light they emit, here determines the speeds and sizes of the solitons that are contained in any wave which begins with an arbitrary shape and develops according to the KdV equation as time goes on: eventually all that will remain are the solitons and some minor oscillations that die out, just like Russell's description.

After their initial discovery in the KdV equation, solitons were found among the solutions of many other nonlinear equations with applications to a large variety of physical and biological systems, and, together with the study of chaos, they have contributed to the re-emergence of nonlinear mathematics as an important field for physics. The dominance of linear methods in physics, which held sway for about seventy five years because of a preoccupation of physicists with the study of electromagnetic radiation and quantum phenomena, is now waning, and the development of the high-speed computer is largely responsible for this.

There is a serious lesson in the soliton story. The discovery of soliton solutions of the KdV equation by computer experiments was astonishing and important, but without an analytical "explanation," that discovery would have been ultimately insignificant and of little intellectual impact. Apart from straight "number crunching," the computer often serves as a very stimulating suggestive tool for discoveries, but it is unlikely to supplant the need for "brain-based" mathematical analysis of physical phenomena.

After this journey through the various kinds of wave descriptions appropriate to different circumstances, and our examination of what they have in common or how they differ, let us return now to a subject first encountered in Chapter 4—Einstein's special theory of relativity. In the next chapter we are going to look into some of its strange consequences and their ramifications.

6

TACHYONS, THE AGING OF
TWINS, AND CAUSALITY

Suppose you play catch in an airplane; would you throw the ball differently? Would the way you throw it depend on the speed of the plane? In other words, are the laws of physics different when experiments are done in a laboratory at rest and when they are done in a moving laboratory? This question had a profound influence on the development of physics in the twentieth century. Let us look at Newtonian mechanics for an answer.

Assume that $\vec{q}\,(t)$ is the position of an object at the time t, $\vec{v}\,(t)$ is its velocity, and $\vec{a}\,(t)$ is its acceleration as observed in the laboratory at rest; and that $\vec{q}\,'(t)$, $\vec{v}\,'(t)$, and $\vec{a}\,'(t)$ are the corresponding quantities for the same object as observed in a laboratory that moves with the constant velocity $\vec{V}$, say, in an airplane.* Then the velocity seen in the moving laboratory is the difference between the two velocities, $\vec{v}\,' = \vec{v} - \vec{V}$. (If $\vec{V} = \vec{v}$, the object is standing still, as seen in the plane.) This means that for each component of the velocity, for example for the x-component, we have $v'_x(t) = v_x(t) - V_x$, and so the plot of $v'_x\,(t)$ looks exactly like that of $v_x(t)$, except that it is shifted

---

*Note that the primes, and the subscripts used below, do not denote derivatives.

vertically downward by the amount V_x (see Chapter 2); similarly for the other two components.

Consider now Newton's law of motion, $\vec{F} = m\vec{a}$, in which m is the mass of the object, $\vec{a}$ is its acceleration, and $\vec{F}$ is the force acting on it, as we have already discussed in Chapter 2. Recall that $a'_x(t)$ is the derivative of $v'_x(t)$ and thus the slope of the tangent, and the tangent of the curve $v'_x(t)$ is parallel to that of the curve $v_x(t)$, because they merely differ by a vertical shift. Therefore, the two accelerations are the same, $\vec{a}'(t) = \vec{a}(t)$: whereas the velocities seen in the two laboratories differ by $\vec{V}$, the accelerations measured are the same. It follows that if the forces observed are equal, the Newtonian equation of motion is the same in both laboratories; it is "invariant."

If the equation of motion is the same in both laboratories, we may conclude that an object that has the same initial conditions (see Chapter 2) will move in the same way in both laboratories. We play catch in the plane just as we do on the ground. There is, in fact, no observable difference between the two laboratories. This is called the Newtonian *principle of relativity*. Its general statement is that there is no experiment using the Newtonian laws of mechanics alone that would allow us to determine whether our laboratory is at rest or in uniform rectilinear motion. Therefore, even though Newton himself had a strong predilection for a God-given absolute space, from the point of view of mechanics, "absolute rest" is a concept that makes no sense; all that is meaningful is the *relative* motion of two observers. Another way of saying the same thing is that there is no mechanics experiment or mechanical measurement performed entirely inside our laboratory that allows us to determine its "absolute velocity."

What about Light?

The nineteenth century, as we discussed in Chapter 4, saw the development of the physics of electricity and magnetism and its theoretical formulation in Maxwell's equations. Light was, in that well-established theory, regarded as a wave that moved in a conjectured all-pervading substratum called the ether, and it moved with a fixed speed. (The Dutch astronomer Olaus Römer had already measured this speed in 1675 by noting the discrepancy between the observed timing of eclipses of the moons of Jupiter and their theoretically calculated timing.) It was, in this respect, similar to sound, which was known to propagate with a speed that is characteristic of the medium through which it moves. The question, therefore, arose whether the absolute speed of an observer, though not measurable by any mechanical means, should not be detectable, after all, by electromagnetic procedures. Could we not measure our absolute velocity, that is, our velocity with respect to the ether, by measuring the speed of light directly in our laboratory, rather than by astronomical means?

As we discussed already in Chapters 4 and 5, if we are moving with a certain velocity $\vec{v}$ through the ether, and we measure the speed of a light ray that moves with the speed c through the ether in the same direction, then we should observe that light ray to have the speed $c - v$; if it moved in the opposite direction, it should be observed to have the speed $c + v$. The difference between the two observed propagation speeds should therefore allow us to determine our absolute velocity, and the Newtonian principle of relativity, though valid in mechanics, could be violated by means of electromagnetic measurements.

This idea, simple enough in principle, was very difficult to implement experimentally, as we have also discussed in Chapter 4, because the speed of light is so enormous when compared to the speed with which the earth could be expected to be hurtling through space. In order for the result to be useful for extracting our own velocity, the speed of light would have to be measured with extreme precision, much greater than was technically feasible. (In addition, it could not be taken for granted that we know the direction in which the laboratory moves with respect to the ether.)

The trick employed by the two experimenters Michelson and Morley in their attempt at performing this difficult experiment was not to measure these speeds, first in one and then in the opposite direction separately, but instead to measure directly their difference, which could be done much more accurately. As explained in Chapter 5, sinusoidal waves such as light exhibit the phenomenon of interference. If two wave trains of the same frequency but different phases are superposed, they interfere with one another either destructively or constructively, or somewhere in between. The result can be seen or photographed as light and dark bands called *interference fringes*, and the distance between the fringes is proportional to the phase difference. For two rays that come from the same light source, that phase difference, on the other hand, is proportional to the difference in travel time that it took for the two waves to arrive at the location of the photographic plate or the observation screen, and this time gap is proportional to the difference in the reciprocals of the speeds of the rays. Therefore, the interference fringes can be used directly, and with great precision, to measure the difference in light velocities required. This was the principle on which Michelson and Morley's "ether-drift experiment" operated, with additional clever refinements. They performed their experiment in 1887 in the city of Cleveland with elaborate precautions and preparations, designed to eliminate mechanical vibrations from the city traffic that might smear out the sensitive detection of the light fringes, which were expected to shift as the entire large apparatus was slowly rotated.

The result of this carefully prepared effort was a huge disappointment. The fringes did not budge! Are we at rest in the ether? Was the ancient notion that the earth is at the center, with the sun revolving around it, right after all? That surely could not be the correct explanation of the "null result" of the Michelson–Morley experiment; there were too many astronomical and physical observations against it. Nor could any of the other explanations offered be satisfactory; all conflicted with well-established experimental results.

Einstein's Theory of Relativity

The only explanation that was ultimately accepted as fully satisfactory by the community of physicists (though there are still a few doubtful nonphysicists) was given eighteen years later by Einstein. It is an interesting psychological fact that Einstein was led to his special theory of relativity, as it is now called, not by the negative outcome of the Michelson–Morley experiment but by imagining, as he had already done as a schoolboy, "riding along on light waves." Indeed, he did not even know about that failed attempt at measuring the ether-drift before he published his theory in 1905, his analogue of Newton's *anno mirabilis,* in which he published three revolutionary papers. (One of them introduced light quanta, the second gave a molecular explanation for the Brownian motion of minute dust specks in a liquid, and the third announced the theory of relativity. The first was later rewarded by a Nobel Prize; the theory of relativity was not mentioned in its citation, nor was Brownian motion. He himself regarded only the first as revolutionary.) Nevertheless, the null result of Michelson and Morley played a crucial role in persuading other physicists of the merits of relativity (a name Einstein later regretted having attached to his theory), the counter-intuitive consequences of Einstein's theory notwithstanding.

Let us now abandon history and look at the basis and some of the consequences of the special theory of relativity, which rests on two fundamental postulates: (1) the principle of relativity, and (2) the constancy of the speed of light.

The principle of relativity is a generalization of the Newtonian principle of relativity, which we discussed above. Instead of being limited to mechanics, however, it is taken to be valid in all of physics. It says that there should be *no experiment whatsoever* that would allow us to determine our absolute velocity. In other words, if identical experiments are performed in two laboratories that are in uniform rectilinear motion with respect to one another, they will have the same results; both laboratories are, in all respects, completely equivalent to one another.

The assumption of the "constancy of the speed of light" may be interpreted as being no more than taking the null result of the Michelson–Morley experiment (which has, in the meantime, been repeated a number of times with even greater precision and always with the same negative outcome) seriously and at face value. It means that for all observers, the speed of light, usually denoted by the letter c, is always the same. If we move along with a light ray, almost with the same speed c, we will still find it to have the speed c relative to us.

It is clear that the implications of this apparently nonsensical supposition have to contradict some of our intuition. Therefore Einstein started out by examining the most basic assumptions required for physical measurements, accepting only the minimum. Each laboratory is presumed to be equipped with measuring tapes or a grid that permits us to ascertain the coordinates of every point in space in it, and with a network of synchronous clocks, so that we can tell the time at which an event occurs, no matter where it happens, without relying on moving one clock from place to place. If we avail ourselves of light signals, which we assume always to have the same speed, it is not difficult to synchronize two clocks at different locations, and to make sure that the rates at which they go are equal. Once such a grid of coordinates and clocks has been established in any given laboratory, the "space-time" of any event, that is, the spatial coordinates of where it happened and the time when it happened, can be pinpointed there.

What we want to find is the "transformation law" that relates measurements of the spatial coordinates and time of a given event in one laboratory to those of the same event in another laboratory that moves, relative to the first, with a constant uniform speed. That transformation law, we shall see, will require some revisions of our "naive" notions. If we assume that the x,y,z axes of the two laboratories are parallel to one another, that the second laboratory moves with respect to the first with the speed V along the x-axis, and that in both laboratories the clocks are set so as to show "zero" when both coordinate origins coincide, then the old-fashioned, intuitive "Galilean transformation" simply says that the y and z coordinates of an event observed in both laboratories are the same, the times are the same, and the x-coordinate x' measured in the moving system is related to x measured in the system at rest by $x' = x - Vt$. This transformation then leads to the commonsense result that the speed of an object or signal that moves along the x-axis with the speed v as determined in the laboratory at rest is seen to move with the speed $v-V$ in the moving laboratory. This conclusion, however, when used for a light signal, conflicts with the postulate of the "constancy of the speed of light." The Galilean transformation, therefore, has to be abandoned.

The Lorentz Transformation

A straightforward exercise using only the two assumptions mentioned (and a minor technical one) can determine the new transformation law that incorporates them; it is called the *Lorentz transformation*. (It is not called the Einstein transformation because the Dutch physicist Hendrik Lorentz had discovered this transformation somewhat earlier, but he failed to give it the new physical interpretation that Einstein advanced.) Rather than writing this transformation down in algebraic form, I will show it to you geometrically, and we will explore some of its consequences. Some of these consequences may appear very strange and utterly counter-intuitive, as they did to everyone at the beginning. Although the theory of relativity has been confirmed by many experimental observations and has long ceased to be controversial among physicists, there are still some nonphysicists who passionately refuse to accept its apparently nonsensical ramifications, and every university physics department periodically receives letters from cranks who think they have found logical flaws in it. For several decades it was also under political attack, particularly in Germany, as a symptom of a disease called "Jewish physics." If such opinions still exist, they are at least no longer openly expressed.

The first thing to be noted about the Lorentz transformation is that it is no longer true that the times shown by clocks in the two laboratories are necessarily equal; the transformation changes all four of the quantities, the time as well as the three spatial coordinates. We must therefore think of an event as existing in four-dimensional space-time. This four-dimensional space-time is usually called *Minkowski space,* after the Lithuanian mathematician Hermann Minkowski. For the sake of visualization, let us ignore the y and z-coordinates of all events and pay attention only to their x-coordinates (as though they occurred along a line rather than in three-dimensional space). Figure 32 shows a coordinate system in which the time coordinate is plotted along the vertical axis and the spatial coordinate along the horizontal axis, but in the form of x/c, where c is the speed of light; we shall denote x/c by X. This simply means that we measure distances X in time-units, like astronomers who measure distances in light-years, which is the distance a light signal travels in one year. We may want to measure times in seconds and distances in light-seconds. A point in this system denotes an event, with a spatial coordinate and a time. If we draw a line parallel to the X-axis through the point, the inter-section of that line with the t-axis gives us its time; to find its X-coordinate, we draw a line parallel to the t-axis through the point, and the intersection of this line with the X-axis determines the X-coordinate of the event.

In this Minkowski diagram, the motion of an object is plotted as a curve that tells us where it is located at every instant of time. Such a curve is called the object's "world line." If the world line is straight, the object is traveling at a uniform speed; the larger the speed the more the line slants away from the vertical axis. The world line of an object at rest is straight up; if it moves with the speed of light, the world line slants at 45°; if it moves more slowly than light, the world line makes an angle less than 45° with the vertical, and if it moves faster than light, it slants more than 45°. A horizontal line connects all events that occur simultaneously; it would be the world line of an object that moves with infinite speed. We may therefore regard the X-axis of this coordinate system as the locus of all points that are simultaneous with the origin, that is, its "line of simultaneity," and the t-axis as the world line of that origin, which is at rest in the system.

The characteristic property of a Lorentz transformation from the laboratory L, in which we denote the spatial coordinate by X and the time by t, to another laboratory L', in which the spatial coordinate is denoted by X' and the time by t', is that the quantity $X^2 - t^2$ has the same value in the two laboratories; in other words,

TACHYONS AND CAUSALITY

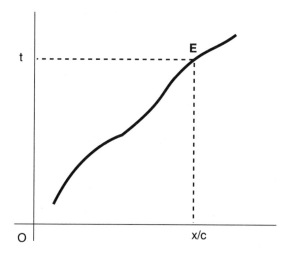

FIGURE 32. The world line of an object in Minkowski space; E is
an event that occurs at the time t at the point x in space.

$X^2 - t^2 = X'^2 - t'^2$. As a result, a light signal, whose world line in
system L is given by the equation $X - t = 0$ if it is sent from the
origin O, is seen to have the equation $X' - t' = 0$, and hence has the
same speed c in L': the speed of light is invariant. This has some
important consequences.

Take a stick at rest in the moving laboratory L', so that it moves
with the speed V in L. In order to measure its length as seen in L, we
must determine the location of its two ends at the same time. Suppose
in L its ends at the time $t = 0$ are at O and x; then in L' one end is
at O at the time $t' = 0$ and the other end is at x' at the time $t' = T$.
Therefore, by the above equation we have $X^2 = X'^2 - T^2$, which
implies that x is less than x'. Consequently, the moving stick appears
to be *shorter* than when at rest! The Lorentz transformation implies
that moving objects are contracted in the direction of their motion;
this is called the *Lorentz contraction*.

Let us next take a clock that is at rest in L'; we want to determine
if its time is correct as seen in L. In order to do this we compare its
time t' to the time t shown by the synchronous clocks at rest in L as
it passes them. Suppose that the location of the clock in L' is
$X' = 0$, and X in L. Then, by the equation given above, we have
$-t'^2 = X^2 - t^2$, or $t'^2 = t^2 - X^2$, which implies that t' is less than t.
Therefore, the traveling clock appears to be slow! This effect goes by
the name of *time dilatation*.

Some other surprises await us in the Minkowski diagram. Figure 33 shows the effect of a Lorentz transformation. The X'-axis of the moving laboratory L' makes an angle a with the X-axis of the laboratory L at rest, and so does the t'-axis relative to the t-axis. The angle a depends on the speed V of L' in the laboratory L; if that speed is small compared with c, then a is small; if V is close to c, then a is close to 45°.

What is the meaning of this diagram? Consider an event E. As seen in L, its time t and its location X are obtained as explained earlier and indicated in the figure. As seen in L', however, its time t' and location X' are determined by drawing lines parallel to the t' and X'-axes and finding their intersections with these axes, as in the figure.

Ponder for a moment what the implications are. The X-axis is the locus of all events that occur simultaneously with the event O at the origin, as seen in the laboratory L; the X'-axis, on the other hand, is the locus of all events that happen simultaneously with O as seen by an observer riding along in L'. Therefore we have to conclude that two events which appear to be simultaneous to one observer do not appear to be simultaneous to another observer who is moving with respect to the first! The concept of simultaneity is no longer absolute, but observer-dependent; clocks that are considered synchronous with one another in one laboratory do not appear to be so when seen from a moving laboratory.

We can now resolve a certain puzzle connected with the Lorentz contraction. The shrinkage has to be perfectly symmetrical; just as an observer in L sees a stick in L' contracted, so an observer in L' sees a stick at rest in L contracted by the same amount. If you think that there is a contradiction between these two observations, because putting the two relatively moving sticks next to one another for an instant should allow an objective determination of who is right, you forget that the observer in L must compare the ends of the stick in L' *simultaneously* by his clocks, whereas the observer in L' compares the ends of the stick in L simultaneously by clocks at rest in L'. The first of these measurements appears to the investigator in L' to measure the ends of his stick at different times, and so does the second seem to the one in L. Therefore a contradiction is avoided, because what is simultaneous in one system is not so in another.

Similarly for our conclusion that a moving clock appears to be slow. Observers who travel with the clock in L' will, of course, also see any clock in L to be slow; both sets of observers see the others'

TACHYONS AND CAUSALITY

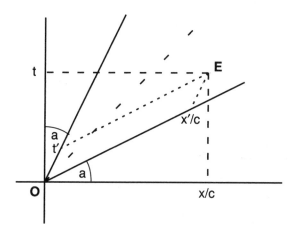

FIGURE 33. The time and space axes of a moving system relative to those of a system at rest; E is an event, seen at t and x in the system at rest, and seen at t' and x' in the moving system.

clock as slow! How is that possible? The answer to this conundrum is that when observers in L say that the clock in L' is slow, they are comparing the moving clock with a string of clocks at rest and sychronous in L, and the clock keeps falling farther and farther behind as it travels along. Both sets of observers must agree on that; but observers in L' will explain it by arguing that the clocks in L are not synchronous. Such observers will also see any individual clock in L falling more and more behind as it passes a string of clocks synchronous in L' and conclude that, as far as they are concerned, the clock in L is slow. There is perfect symmetry between the two sets of observers. The amount by which a moving clock slows down, of course, depends on its speed in comparison with the speed of light, and it is noticeable ordinarily only if that speed becomes comparable to c.

The slowing down of moving clocks is independent of the mechanism of the clockwork; it applies to all natural phenomena and is experimentally well verified. There are, for example, many particles in nature that tend to decay, like a radioactive nucleus. While the exact time at which such a particle decays cannot be predicted, as we will discuss in Chapter 7, each has a well-determined "half-life" T; this means that if we have a collection of them, half will have decayed after the time T. Suppose we produce a beam of such particles in an accelerator, all of them traveling at the same high speed close to the speed of light. We then find experimentally that their half-life is much longer than T by an amount that agrees exactly with the prediction of the time dilatation from the Lorentz transformation.

The Traveling Twins and Superluminal Signals

As I have emphasized, the slowing-down effect of clocks is a perfectly symmetrical affair: I see your clock as slow, and you see mine as slow. There is, however, a way of destroying that symmetry. Imagine a pair of twins, Julie and Rachel; at the age of thirty, Rachel embarks on a long space journey in a very fast spaceship while Julie stays behind. After ten years, Rachel returns to Julie, who is now forty years old; but Rachel is only thirty five! I don't mean by that that Rachel has somehow seen only five summers to Julie's ten, but that she has biologically aged in every respect by five years, while Julie has aged ten years. This is sometimes called the "twin effect" or the "twin paradox" because it appears to violate the symmetry between the two clock slow-downs. The reason it implies no such violation is that in order to return to her twin, Rachel must have been accelerated and decelerated in some fashion, whereas Julie has not. It is this lack of constancy in the velocity of one of the clocks that accounts for the left-over asymmetry. There is, indeed, a twin effect, but it is no paradox.

Let us now return to our Minkowski diagram and consider the event E indicated in Figure 34; it lies below the 45° line. As seen in laboratory L, this event takes place at a *positive* time, that is, it occurs after the event O at the origin because it lies above the X-axis, which is L's line of simultaneity through O. As seen in laboratory L', however, the same event takes place at a negative time because the line through it parallel to the X'-axis intersects the t'-axis below O, and so it lies below the line of simultaneity through O as seen in L'. As we see, the order in which the two events O and E occur depends on the observer.

Imagine now that we in L send a signal from O to E; since E lies below the 45° line, this signal would have to be "superluminal," in other words, it travels faster than light. That it was sent from O to E means that O occurs before E. However, if we look at the same signal from the vantage point of an observer in L', the event E occurs before O; such an observer would therefore conclude that the signal was sent from E to O. (If the signal traveled more slowly than light, E would lie above the 45° line, and no laboratory would exist in which the time of E is negative.) Would this not be extremely confus-

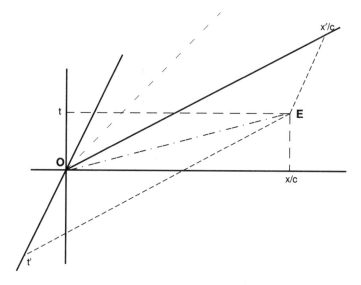

FIGURE 34. Another event E seen in two laboratories. The dot-dashed line represents a superluminal signal connecting O to E.

ing and wreak havoc with our notions of causality? Surely we must be able to distinguish between the person who caused the signal to be sent and the passive receiver. But such a disturbing confusion of causality would certainly arise if any kind of signal that travels faster than light were possible. For any such signal a laboratory would always exist, traveling like L' in L more slowly than light, in which the signal would appear to travel from its receiver to its sender. For this reason, it is often asserted that the theory of relativity forbids the existence of superluminal signals, or that it forbids anything to move faster than light. We should remember, however, that the prohibition of such speeds comes from a combination of the theory of relativity and our ideas of causality, and not from relativity alone.

The Adventure of Two Spaceships

Perhaps this thought experiment will drive the message home. Figure 35 shows the world lines of two rocketships, both initially at rest (their world lines are vertical) and subsequently traveling with the same velocity, less than the speed of light. At a certain moment, shown by A, the passengers of rocketship RI use their superluminal transmitter to send a message to rocketship RII. As is visible in Figure 33, the world line and the line of simultaneity of a uniformly moving

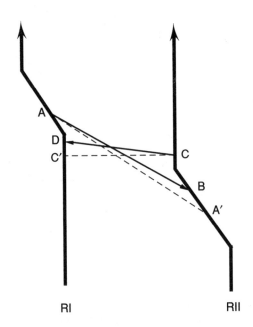

FIGURE 35. The world lines of two rockets sending messages to one another via superluminal signals; the dashed lines are the lines of simultaneity.

system are reflections of one another across the 45° line. Figure 35 shows the line of simultaneity of rocketship RI from the point A to the point A'. (Since the two world lines of the rockets at A and A' are parallel, this is also the line of simultaneity of RII at that time.) The message arrives a little bit later at B. The passengers of both rockets agree that B is later than AA', their mutual line of simultaneity; they therefore agree that the message was sent from rocketship RI to RII, and there is no confusion.

At a later time, after RII has stopped, its passengers send a superluminal message back to RI, at C. Their line of simultaneity is now horizontal and intersects the world line of RI at C', where the receiver is also at rest. The message arrives a little later at D; again both agree that D is later than C and therefore, as far as both are concerned, the message was sent from RII to RI.

But now look at the astonishing result: *For rocketship RI the return message arrives at D before the original message was sent at A.* In effect, by means of the round-trip superluminal signal from RI to RII and back, the passengers of the first rocket managed to send a message into their own past. They could tell themselves what will

TACHYONS AND CAUSALITY

happen at a later time; it is a clear case of precognition. Or you could imagine that the message sent was to ignite a bomb at D to blow up rocketship RI, so that the message at A that set off the explosion could not have been sent. The resulting "causal cycle" has landed us in a contradiction.

What are we to make of this? We must recognize that the "contradiction" we arrived at is not a *logical* contradiction. After all, it is conceivable that when the passengers tried to send a message at A to blow up their own rocket at D, the apparatus simply would not work no matter how often and hard they tried. The contradiction would then be averted; but it would be at the price of something we have never experienced, that certain kinds of messages simply cannot be sent. In spite of frequent glitches and malfunctionings we all know about, on the basis of a vast amount of past experience we expect, ultimately, to be able to send any message we like. On the grounds of that experience, we do believe in *causality;* and it is this belief in causality, together with the theory of relativity, that would be violated by the existence of superluminal signals.

Causality

We must conclude that the combination of relativity and causality implies that the speed of light is an upper limit to all possible speeds, in some sense. Are there speeds that are exempt from this prohibition? We can see from Figure 33 that the Lorentz transformation implies, right from the start, that no two laboratories can move, relative to one another, with a speed greater than c. The Minkowski diagram would clearly become all mixed up. We can also demonstrate that under a Lorentz transformation the three components of the momentum vector $\vec{p}$ of a particle and its energy E behave just like the three components of the position vector and the time. It turns out that if a particle could move with a speed greater than c, it would have to have a mass m such that m^2 is negative, which would imply that m is an imaginary number (see Chapter 1). Even to accelerate it up to the speed of light would require an infinite amount of energy. Therefore, certainly ordinary particles must always move with subluminal speeds. In addition, there is, of course, the argument that if they could move with superluminal velocities, they could be utilized to send messages that would violate causality. No superluminal signals are allowed.

But what constitutes a signal? It is clear from our arguments above that a signal from A to B is anything that can activate a switch at B on command from A. This would necessarily include any transmission of energy or of information. However, there are some motions that are not of that nature. For example, if you hold your finger in front of a lightbulb and observe the shadow cast on a distant screen, the speed of the shadow, as you move your finger, is proportional to the distance to the screen. If that distance is large enough, in principle the shadow can move as fast as you please. However, as it rushes from a point A to a point B, both on the screen, it cannot be used to activate a switch at B on command from A. Therefore, a superluminal speed of such a shadow would not violate the relativity-causality based prohibition. There are many other speeds of phenomena in physics that are also free from this limitation.

The joker in the argument of the last paragraph is the phrase "on command from." It may stimulate you to contemplate the role that our belief in "free will" plays in what we mean by causality. We certainly believe that we are free to send any message we want, and to observe its effect. Indeed, all scientific experimentation is based on the assumption that we are free to wiggle something here and to record the response over there. If we are not free, and our wiggling is determined by what we regard as its distant effect, the conclusions we draw from any experiment are invalidated. Nor could we escape from such problems by the substitution of a random device for our free will because, unless we take it for granted that causes must always precede their effects, for which we have plenty of evidence but which is not logically required, even the throw of a pair of dice may be determined by the later events that we regard as its "effects." These thoughts should not be interpreted as serious doubts about our ideas of causality; they are merely meant to show that these notions are not based on logic alone, but on experience, and a vast amount of it.

The prohibition of superluminal signals is not the only place in physics in which the concept of causality is employed, either explicitly or implicitly, to draw physical conclusions from a theory. Such a use of causality is, in fact, not uncommon; the following is another good example. The Maxwell equations imply mathematically that any accelerated moving electric charge is accompanied not only by the Coulomb field (that is, the field produced by an electric charge at rest, as described in Chapter 4) but by an additional electromagnetic field that dominates the Coulomb field at large distances. This field, however, is not uniquely determined by the equations; we also need an

TACHYONS AND CAUSALITY

initial condition. Such an initial condition is provided by the plausible assumption that there was no such field present before the charge was accelerated; if the acceleration *causes* the field, it should be observable only later, and not before the acceleration took place. Any such field that existed earlier might effect the acceleration by exerting a force, but it could not be the result of the acceleration. The particular solution of the Maxwell equations that is selected by this initial condition is called *retarded*. Other solutions of these partial differential equations are sometimes useful for mathematical purposes, but they are usually assumed not to be of any direct physical use. Based on our universal experience that causes precede their effects, we assume that only the retarded field is of physical interest. We can then calculate from Maxwell's equations, together with the retardation condition, exactly how much electromagnetic radiation is produced by an accelerated electric charge, and the result is confirmed by detailed experiments and observations.

Tachyons

As we discussed in Chapter 2 and will explore in more detail in Chapter 7, the advent of the quantum theory severely damaged our conviction in the universal sway of causality, at least in the old-fashioned sense of that word. Whereas the assumption of strict determinism used to be regarded as a *sine qua non* of the scientific mode of thinking, or indeed as a necessary "Kantian" concomitant of rational thought, we are now accustomed to thinking of causality as based on experience and observation, as propounded by the philosopher David Hume. It therefore, not surprisingly, occurred to some physicists that the arguments based on signals, causality, and relativity might be invalidated by quantum mechanics. Since the motion of a particle cannot be precisely predicted, perhaps there was no reason, after all, why a particle could not exist that moves faster than light. Such a particle was called a "tachyon" from the Greek word for "swift." It would have to have an imaginary mass, as I pointed out above, but so what?

Tachyons would have other strange properties: because they could never be accelerated from rest up to and past the velocity of light, they would *always* have to rush around at superluminal speeds; as they lose energy, say, by radiating light if they are electrically charged, they speed up. It would take an infinite amount of energy to slow a tachyon down to the speed of light. Well, such properties are strange, but physicists are inured to the strangeness of the natural world. So several experimental physicists decided that, even if most theorists strongly believed that such particles could not exist and the chances for success were extremely small, it was worthwhile to search for tachyons. (An experimental scientist worth his salt should always spend more of his efforts trying to *disprove* the constructions of his theoretical colleagues than attempting to prove them. The question of doing an experiment with unlikely success but high pay-off, as against one that is likely to succeed but would not make much of a splash, is the same question that has to be weighed by any gambler; it is answered mostly by temperament.) To no one's great surprise but to the regret of some, tachyons were not found. At the present time, there are no active searches for them.

We are left with the conclusion that, if the theory of relativity is correct, and there is certainly a vast amount of experimental evidence in its favor, and if, in addition, our notions of causality, for which there is also a large body of evidence, are not off base, we cannot expect to receive messages that were dispatched only a few weeks ago from aliens who live on a planet circling a star that is 100,000 light years away; no matter how such messages reach us, they must have been sent at least 100,000 years ago. On the other hand, if aliens from that planet came to visit us by superfast spaceship, they may have experienced the trip as having taken no longer than a few months, and if, in some future generation, space-faring voyagers embark on a ten-year exploration in a high-speed spacecraft, they may find, upon their return, that the earth they left behind has aged a hundred years.

In this chapter we have discussed one of the two revolutionary theories of physics developed during the first half of this century, Einstein's special theory of relativity, and some of its astonishing consequences. We now turn to the other revolution in physics during to which Einstein also made seminal contributions but which, in the end, he disavowed—the quantum theory. As we will see, its consequences must alter our perception of the way nature works even more profoundly than relativity.

7

SPOOKY ACTION AT
A DISTANCE

We have already seen in Chapter 4 that the quantum theory, developed during the first quarter of this century, brought with it momentous transformations in the way we understand nature. For the most part, these changes originated in the abandonment of strict determinism, not only for long periods of time as in classical mechanics but, in the atomic and nuclear world, even for short times and at the most basic level. There are historians of science who would hold the irrational, antideterministic intellectual atmosphere and the political chaos in Germany after its defeat in World War I responsible for the emergence of quantum mechanics. "Without the Munich of 1919," that is, the ultimately unsuccessful revolution setting up a Bavarian Soviet Republic and the resulting counter-terror, argued Lewis S. Feuer, "Werner Heisenberg would not have conceived the principle of indeterminacy." Such provocative external sociological explanations of the structure of science sometimes contain enough of a grain of truth to sound plausible, but they distort the nature of science. To account, in part, for the psychological origins of new ideas is insufficient to explain their experimental corroboration, their wide acceptance, and their long staying power.

The quantum theory, however, implies an even deeper and more far-reaching modification in our concepts than the destruction of strict causality: we must reconsider what constitutes "reality." We shall examine both of these aspects of the quantum theoretical revolution by beginnning with a brief look backwards to the *ancien régime*. In Chapter 2 we explored the structure of classical Newtonian mechanics as formulated in various ways by the great physicists and mathematicians of the last 300 years, and in Chapter 3 we discussed the ideas developed by Maxwell, Gibbs, and Boltzmann in the second half of the nineteenth century, leading to statistical mechanics. Let me summarize some of their main characteristics.

Classical Mechanics and Statistical Mechanics

The state of a physical system (such as a system of rigid bodies or of particles) at a given time is defined by the positions (and in the case of rigid bodies, orientations) and velocities (or momenta) of all of them. Although this involves an idealization, in principle the state of a system is assumed to be experimentally ascertainable, that is, measurable with arbitrary accuracy. There are, of course, practical, technical, limitations to the precision of such measurements, but in principle advances in technical means can be expected to bring unlimited improvement to the possibility of our knowledge.

The laws of classical mechanics are such that, if we know all the operating forces and the state of a system at an arbitrary time t_1, it is possible in principle to determine, or calculate, the state of the same system at any later (or earlier) time t_2. Thus the universe runs like a perfect clockwork, and its state at any time is completely determined by its state at any earlier time. This was the basis of Laplace's famous claim, quoted at the beginning of Chapter 2, that, given exact knowledge of the state of the universe at the present time and sufficient intelligence, we could precisely predict the course of the world into the indefinite future.

In Chapter 2 we saw that there are certain deep practical limitations to this expectation, and these are not just the obvious ones that we cannot know the present state of a physical system with unlimited accuracy and are therefore unable to predict its future course precisely. In fact, for most systems, a small error in our present knowledge will make predictions far into the future essentially impossible. Because of sensitivity to initial conditions, tiny initial perturbations

144

usually lead to eventual chaos. Thus the most powerful computer cannot make even approximate predictions for an indefinite future.

Such practical limitations on the predictability of complicated mechanical systems become, of course, more and more severe as the number of particles involved increases. The calculation of the motion of the planets in the solar system, when all their influences upon one another are taken into account, is a formidable task. Trying to do the same for all the molecules in a container full of hydrogen gas would be horrendous. Even the most powerful computer would find it impossible to track each of some 10^{23} particles.

The field of physics that deals with large numbers of objects—such as are contained in gases and fluids—and with the laws that govern their properties (such as temperature, pressure, volume, and so on), is thermodynamics, which we examined in Chapter 3. This discipline is a beautiful, self-contained subject whose structure was developed in the nineteenth century without any use of underlying models or explanations based on more fundamental theories. At that time, after all, the atomic theory of matter was not well established and, indeed, was still quite controversial. The first and second laws of thermodynamics, on the other hand, were well established and accepted.

As the existence of atoms and molecules became more and more universally recognized, the need arose to "explain" the laws of thermodynamics in terms of the behavior of the constituents of the fluids and gases, that is, the atoms and molecules. The behavior of these constituents, of course, was understood to be governed by the classical Newtonian laws of motion. Therefore, the laws of thermodynamics ought to be reducible to these more "basic" laws; the main fact connecting the two sets of laws was the enormously large number of particles in any container utilized in the laboratory. This gave birth to the science of statistical mechanics. The laws of thermodynamics, originally autonomous and independently discovered and formulated, became secondary manifestations of the more fundamental laws of mechanics, thus losing their apodictic, "necessary" character (if the Newtonian laws are accepted) to a statistical one. The second law no longer stated that heat *never* flows from a cold to a hot body; it now said that such a heat flow is extremely improbable. This, of course, is a major change in principle, but it has no practical consequences because, owing to the large number of particles involved, occurrences that were formerly "forbidden" had such small probabilities that only a few, if any, could be expected even during the entire age of the universe.

The language in which statistical mechanics was formulated incorporated the idea that a physical system could be duplicated many times and allowed to run its course, subject always to the same laws and constraints, but possibly from different initial conditions. That is, the molecules might start from different initial positions or velocities, but from a macroscopic perspective we would be unable to tell the difference. Gibbs called such large collections of identical physical systems *ensembles*. Probabilities are then calculated as ratios of occurrences in such ensembles, just as the probability of getting "heads" in a coin toss may be calculated as the fraction of the number of heads that will be found in an ensemble of tossed identical coins. In principle, an ensemble should have infinitely many members, but in practice, "very many" will suffice. (We will ignore here the deep and subtle mathematical problems this raises.)

The situation in physics at the beginning of the twentieth century was, in sum, such that the future behavior of many physical systems like the planets and their moons was thought to be exactly predictable if only the present state of these systems were known accurately (which was an idealization that could *in principle* be approximated arbitrarily closely). The behavior of other systems, consisting of very many molecules, could, for practical reasons, be predicted only statistically, but no one doubted that even in the microscopic realm "determinism" reigned. In a basic sense, the statistical laws were epiphenomena born of our clumsiness or technical lack of sophistication.

The Quantum Theory

The turn of the century brought a fundamental change to this view with the advent of the quantum theory, several facets of which were already introduced in Chapter 4. Let us look at some especially pertinent characteristics of the new theory. Niels Bohr's model of the atom, which superficially resembled a miniature solar system with the nucleus as the sun and the electrons as the planets, postulated that, contrary to Maxwell's laws of electricity and magnetism, the electrons move around the atomic nucleus in certain "allowed" orbits without emitting radiation. When this was combined with Pauli's "exclusion principle," according to which only two electrons could ever exist in the same orbit (one with its spin pointing up and the other down), it provided a complete explanation of the periodic table of the elements, which Dmitri Ivanovich Mendeleev had constructed ad hoc on a purely empirical basis. In Bohr's model, it is only when an electron "jumps" from one orbit to another of lower energy that light is emitted in the form of a photon of definite energy, corresponding to light of a definite wavelength, that is, color. This light is observed as the *spectrum* of emission by "excited" atoms, for example, in a heated gas, which is restricted to certain characteristic "spectral lines" that account for the difference in the color of street lamps using mercury-vapor and those using sodium. There are several revolutionary ideas contained in this model. The one I would like to focus on is the notion that the time at which this jump occurs, and hence exactly when light will be emitted by the atom, cannot be predicted. All that we can predict is how long on the average the atom will remain in the "excited" state before the jump occurs—the statistical distribution and other statistical details of the jumps, and so on—but not their exact timing.

Similarly for a radioactive nucleus. If a certain kind of nucleus is subject to radioactive decay, giving off, for instance, an alpha particle (a helium nucleus), its "lifetime," that is, the average time it takes before it disintegrates, can again be exactly predicted, and so can all sorts of statistical properties. Exactly at what time an individual nucleus will decay, however, is unpredictable.

This "acausal" nature of the quantum theory, whose initial form Planck, Einstein, and Bohr fashioned early in this century and whose final version Heisenberg, Schrödinger, and Dirac formed in the 1920s, is perhaps its best-known feature. It is one that both attracts many nonphysicists who strongly dislike the deterministic clockwork universe of classical physics and repels many others who find a world that resembles a roulette wheel unacceptable. It, and various of its interpretations, have found their way into philosophical and quasireligious popularizations and have been exploited for illegitimate nonscientific purposes. A number of the very founders of the quantum theory were quite unhappy with its acausal features, which were particularly emphasized by Heisenberg, and they tried to find alternatives, without success. Einstein expressed his dissatisfaction by asserting his disbelief that "God played dice with nature." Let us explore what the origins of this view of the universe as a gambling casino are, but also go further and try to understand the issues in what can only be regarded as one of the truly great intellectual debates of Western thought.

The crux of the matter is that, contrary to many misconceptions, Einstein's objection to the quantum theory during the last 30 years of his life was not so much based on its lack of "causality" as on his disagreement with its treatment of "reality." The great debate consisted of two papers in *Physical Review,* the first, in May 1935, by Einstein and two of his associates at the Institute for Advanced Study in Princeton, Boris Podolsky and Nathan Rosen (this paper is usually referred to as EPR), and the second in October 1935, a reply by Niels Bohr, to whom such an objection from Einstein was a matter of agonizing importance. Both were entitled "Can quantum mechanical description of physical reality be considered complete?" These papers have led to a large number of comments and interpretations by historians, philosophers, mathematicians, and physicists. But the vast majority of physicists have more or less ignored them, either because working physicists have little taste for what they consider "metaphysics," or because, though some may not really have fully understood Bohr's reply, they believe Bohr to have answered Einstein satisfactorily. (Niels Bohr, a great admirer of Kierkegaard, had a characteristically opaque, and sometimes even mystical, style of expressing himself.)*

Suppose we begin by discussing both the quantum theory in general and a philosophical issue that, though important and relevant in itself, does not really get quite to the heart of the EPR problem (or, as some call it, the EPR paradox). We will then turn to the real EPR issue and some recent experiments that have an important bearing on it.

Indeterminacy

In quantum mechanics, the state of a physical system at a given time, if it is known as well as it can be, is described or symbolized by what is called in various contexts either a *state vector,* a *wave function,* or a *wave packet.* This state vector changes in the course of time in accordance with a first-order differential equation (see Chapter 2), known as the Schrödinger equation. Just as in classical mechanics, if we know the state of a system at one time (and the forces on it or the interactions), the state at any later (or earlier) time is completely determined. To every dynamical variable of the system—for example, the position or the momentum of any particle in it—there corresponds an operator, which is simply an object that transforms one

---

*The "EPR effect" has even recently been seriously advocated as a quantum mechanical method of encrypting messages.

wave function into another in a prescribed way. (We have already encountered operators in Chapter 4.) In contrast with classical mechanics, however, knowing the state of a system does not necessarily imply knowing the experimentally observable values of all its dynamical variables. For example, if the wave function of a system consisting of one particle is known, the probability of finding it in a certain region of space is determined, and so is the probability that its momentum lies in a certain interval. The state may be such that its momentum is given with great (but not infinite) precision. In that case, it turns out its position cannot be determined very accurately. Or the other way around, if its position is almost precisely known, its momentum is "smeared out." The product of the two accuracies (with which its position and momentum are determined) must always be larger than a certain constant of nature called Planck's constant, which we have also encountered earlier. Thus, if one is small, the other must be large. This is the famous Heisenberg indeterminacy principle, touched upon in Chapter 4. It sets a specific bound on the precision with which certain pairs of dynamical variables can be simultaneously measured or known. These bounds are not imposed by present technical limitations but, if the quantum theory is correct, have to remain in place forever.

Similar arguments apply to dynamical variables other than position and momentum, for example, to the three components of the angular momentum. (We will discuss the concept of angular momentum in a bit more detail in Chapter 10.) Particles are known to have a certain intrinsic angular momentum, called spin. But though the x, y, and z components of this angular momentum can in principle be measured separately as precisely as we like, they cannot be simultaneously determined with unlimited precision. We can, for instance, prepare a beam of electrons all of whose spins point precisely "up." If we measure their vertical spin components in a second apparatus, we find that they all point up again and none point down. However, if we run that same beam through a device that measures their horizontal spin components, say, in the east-west direction, we find that half the spins point to the east and half to the west. (See Figure 36.) If the part of the beam consisting of particles whose spins point to the east are run again through a vertical measuring device, half of the spins will be found to point up and half down! In other words, among the particles whose spins point up, the probability for one of them to have its spin point to the east is one half, and among those whose spins point to the east, the probability for spin up is also one half.

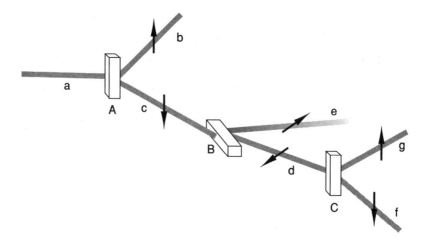

FIGURE 36. Apparatus A tests the particles in beam *a* for spin up or down; those with spin up go along *b*, and those with spin down follow *c*. Instrument B tests those in beam *c* for spin east or west, and those with spin east follow *d*, while those with spin west, *e*. C again tests the particles with spin east for spin up or down, and half are found with spin down and follow *f*, while half have spin up and go along *g*.

This apparently strange state of affairs is not a matter of idle speculation but has been extremely well verified by many experiments. Heisenberg gave an intuitive "explanation" of these peculiarities that has become traditional—and has unfortunately been transferred by some nonscientists to other contexts such as psychology and economics without justification—namely, that the measurement of one spin component necessarily disturbs the other components.* (The disturbance of the value of one variable by the measurement of another is embodied in the fact that the operators that correspond to the two do not commute, as we discussed in Chapter 4.) However, we shall see that this explanation cannot be taken quite seriously.

Incidentally, you may wonder why the experiments I have described ask "Is the spin up?" rather than "In what direction does the

---

*The transfer of the concept of Heisenberg's indeterminacy principle to other fields is usually based on a misunderstanding of its intuitive "explanation." That measurements produce disturbances in the quantity measured is, of course, true in many areas of science; this has no relation to Heisenberg's principle. The "disturbance" at the quantum level is an irreducible effect of the measurement of one quantity on other specified properties. In fact, if the same quantity is measured twice in succession, no fundamental disturbance in the second measurement produced by the first is postulated by quantum mechanics.

spin point?" Just as the most efficient and useful language employed by computers uses only strings of zeros and ones, so the most effective way of setting up experiments with unambiguous results is to ask nature questions that can be answered by yes or no. Other kinds of questions, indeed, may, according to quantum mechanics, have no meaningful answers.

On the basis of these simple examples (and we shall return to the case of the spin angular momentum again), we may infer a characteristic of quantum mechanics: while, just as in classical mechanics, the state of a system at one time determines its state at any later time, the meaning of the word "state" has been altered. Classically, knowledge of the state of a system allows, indeed implies, precise statements about the numerical values of all dynamical variables (or, as some physicists prefer, answers to yes/no questions); quantum mechanically, knowledge of a state implies only *probability* statements about similar questions. There is perfect determinism from one time to another for the states of a system, just as there is in classical mechanics, but knowing the state of a system does not mean the same thing in the two theories.

Interpretations

The question then arises: what exactly is the meaning or interpretation of the wave function, or state vector, in quantum mechanics? There are several different kinds of interpretations that have evolved in time, and even today there is disagreement among physicists, including some of the most prominent ones.

The *realistic* interpretation regards the wave function as a "condition of space," analogous to an electric or magnetic field, as discussed in Chapter 4. This was the earliest view, but for technical reasons that would take us too far afield, it is not really tenable, and few physicists still hold it. Nevertheless, it is an intuitive idea that is almost impossible to avoid in introductory college courses and, subsequently, very difficult to eradicate.

As I indicated before, the state vector (which is the same thing as the wave function) is at least in part determined by the measuring process, and in the realistic interpretation it is a function of physical space and time. If I measure the vertical component of the spin and find it "up," I am at the same time preparing a state for which that spin component is up. This leads to a notorious difficulty, that at the moment of the measurement, the wave packet instantaneously "collapses" in all of space and is reduced to one for "spin up." In a realistic interpretation, such a sudden collapse everywhere is hard to envisage physically.

Among the *nonrealistic* interpretations, the most extreme is the *subjective* one. Its proponents regard the wave function as an expression of knowledge on the part of an observer. The measuring process influences the wave function because it changes the knowledge of the experimenter. "If the packet is to be reduced, the interaction must have produced knowledge in the brain of an observer. If the observer forgets the result of his observation, or loses his notebook, the packet is not reduced," said the American physicist Edwin Kemble. I do not believe that this view has many adherents among contemporary physicists, though it does have some quite prominent followers. The subheading of an article by Bernard d'Espagnat in *Scientific American* reads, "The doctrine that the world is made up of objects whose existence is independent of human consciousness turns out to be in conflict with quantum mechanics and with facts established by experiment." This view of the world has a long history in philosophy under the name of "idealism," and one might be tempted to respond to the modern-day idealists as Samuel Johnson did in the eighteenth century, by kicking a rock to refute Bishop Berkeley's notion that reality was all in our minds. In some countries, the interpretation of quantum mechanics even became an important ideological issue, with strident attacks on "idealistic" interpretations in the Soviet Union under communism, whose official philosophy, after all, was "materialistic." (A well-known textbook on quantum mechanics by D. I. Blokhintsev contained an "idealistic" interpretation in its first edition in 1944 and a "materialistic" one in its second edition in 1949, in which the author strongly rejected the view he had advocated earlier.)

The majority of today's physicists, I believe, use and teach the *objective* nonrealistic interpretation, which was first consistently advanced by the German physicist Max Born: the wave function is a measure of an objective probability. The next question, and again a controversial one, then is, of course, what is the physical meaning of "probability"? There are those who adopt the "frequency theory" and others who believe in what is sometimes called a "propensity theory." In the first theory, the probability of heads in a coin-toss is defined by an infinite repetition of coin tosses (or infinitely many simultaneous ones with identical coins); in the second, it is a "propensity" attached to individual coin tosses.

The difference between the two views of probability is not confined to quantum mechanics and may be illustrated in other contexts. What do we mean by saying, "The probability for Smith to die in 1993 is x%"? Insurance companies depend on an exact calculation of x for the premium they charge Smith, and for their calculation they must assign Smith to a large population (an *ensemble*) for which they have reliable statistics. If all the company knows is that Smith lives in the city of New York and is 45 years old, their actuarial tables tell them what the probability x is, and they can fix a premium. On the other hand, if they also know that Smith smokes and has had two heart attacks, their tables will list a quite different probability. In other words, Smith may be considered to belong to many different ensembles, and the probability of his death depends on which of these populations is chosen. If many members of one of these ensembles are picked, the statistics of their deaths define the probability for each one of them to die in a given year, and the probability that Smith will die in 1993 is completely defined by regarding him as a member of that population. This is the frequency interpretation of probability. In the propensity interpretation, on the other hand, the probability of Smith's death in 1993 is attached intrinsically to Smith as an individual; different choices of populations to which he belongs are merely statistical methods to find out what that inherent probability might be.

Most physicists have their first contact with probability theory in statistical mechanics. They are therefore accustomed to its statistical interpretation in terms of ensembles. As already mentioned in Chapter 3, these are infinite collections of imagined duplicates of a given physical system, all developing in the same way but with different initial conditions. We would consequently expect these scientists to have a natural predilection for the frequency theory of probability (which I personally much prefer). An examination of physicists' writings on quantum mechanics, however, shows that this is not universally true, for there are many who think of probability attached to individual events as a propensity. Probability, in the view of the followers of this interpretation, has a meaning that requires neither an infinite repetition nor an infinite duplication of the system. ("Infinite" is always, of course, meant as an idealization and is to be approximated by "a large number.")

For physicists with different views of the meaning of probability, then, even the objective nonrealistic interpretation of quantum mechanics carries different meanings. The frequency adherent, on the one hand, sees the wave function as always and intrinsically referring to ensembles. The wave function makes statements about individual systems only insofar as they are members of a large set of identically prepared duplicates. The propensity buffs, on the other hand, think of the state vector as attached to an individual system and as testifying to its predilection for certain responses. For the propensity theorists, the reduction of the wave packet by a measurement is a perplexing and mysterious event because it changes an ingrained property of the system; the frequency probabilist sees less of an inherent mystery in it. In the view of followers of the frequency theory, the measurement intrinsically must be done many times on identical systems, and the reduction of the wave packet merely describes a selection process that forms a new ensemble from chosen members of an old one.

The propensity interpretation of probabilty leads its adherents to a view of the quantum theory that has one resemblance to classical mechanics: it deals with individual systems. However, where classical mechanics makes definite predictions of values for dynamical variables, quantum mechanics makes only probabilistic ones. The followers of the frequency interpretation, on the other hand, see the quantum theory as analogous to statistical mechanics: it always deals with ensembles, never with individual systems. From this point of view it then becomes tempting, but by no means inevitable, to ask: for quantum mechanics, what is the analogue of classical mechanics in its role underlying statistical mechanics? In other words, statistical mechanics makes probabilistic predictions based on an underlying deterministic theory, namely classical mechanics. May there not be, similarly, a deterministic theory, so far unknown, that would underlie quantum mechanics? The probabilistic nature of quantum mechanics would then be the result of our ignorance of the underlying structure, just as the statements of thermodynamics are the statistical results of our ignorance of the enormously complicated details involved in systems of very many particles.

Such underlying theories that would "explain" quantum mechanics are usually called *hidden variable theories* because they would introduce additional dynamical variables that remain hidden from us, perhaps even in the most refined experiments. However, in terms of these covert parameters, the outcome of experimental results would, in principle, be completely determined. Over the last sixty years small groups of physicists have persistently tried to construct such hidden-variable theories, all without success. It is in this context that the famous paper by Einstein, Podolsky, and Rosen plays such an important role. Though these authors did not attempt to formulate a hid-

SPOOKY ACTION AT A DISTANCE

den-variable theory for quantum mechanics, they meant to demonstrate, by an example, that quantum mechanics could not be a *complete description of reality.*

The Question of Reality

I want to emphasize that the issue is not whether quantum mechanics is a final physical theory. Most physicists recognize that no theory is final and that all theories will eventually be superseded, just as Newtonian mechanics was superseded by Einstein's relativity theory. The issue here is whether what the quantum theory describes is *reality.* If so, does it describe *all* of reality, and does that description include such weird elements as "spooky action at a distance," as Einstein called it, that is, an instantaneous influence of one event upon another over large distances? (What Newton once took for granted has now become "spooky"!) It was the lack of such an acceptable description of reality at the microscopic level that bothered Einstein much more than the lack of determinism. As he wrote to Max Born, with reference to two space-time regions A and B that are far apart from one another:

That which really exists in B should . . . not depend on what kind of measurement is carried out in part of space A; it should also be independent of whether or not any measurement is carried out in space A. If one adheres to this program, one can hardly consider the quantum-theoretical description as a complete representation of the physically real. If one tries to do so inspite of this, one has to assume that the physically real in B suffers a sudden change as a result of a measurement in A. My instinct for physics bristles at this.

In another letter, he admits that physicists who regard the descriptive method of quantum mechanics as definitive would be correct in pointing out that the theory makes no explicit use of a requirement that objects in regions A and B have an independent reality and that an external influence on one does not directly influence the other. But, he says,

when I consider the physical phenomena known to me, and especially those which are being so successfully encompassed by quantum mechanics, I still cannot find any fact anywhere which would make it appear likely that [that] requirement will have to be abandoned. I am therefore inclined to believe that the description of quantum mechanics . . . has to be regarded as an incomplete and indirect description of reality, to be replaced at some later date by a more complete and direct one.

For Bohr, on the contrary, there was no reality other than that brought about by the experiment itself. In his reply to EPR he wrote:

The extent to which an unambiguous meaning can be attributed to such an expression as "physical reality" cannot of course be deduced from *a priori* philosophical conceptions, but . . . must be founded on a direct appeal to experiments and measurements . . . In fact, this new feature of natural philosophy means a radical revision of our attitude as regards physical reality.

Another giant among the developers of quantum mechanics, Wolfgang Pauli, essentially agreed with Bohr and considered the search for an unknowable objective reality as the analogue of the medieval scholastic question of how many angels could dance on the point of a pin.

The EPR Argument

Let us then get to the famous EPR argument itself, in the somewhat more intuitive form that the American physicist David Bohm gave it. In a nutshell, the idea is to imagine the disintegration of an atom or molecule that has no intrinsic spin into two particles with spin, say atoms or electrons, which then fly off in opposite directions. The conservation law of angular momentum (which we shall examine in more detail in Chapter 10) implies that the spins of the two escaping particles must necessarily be equal and opposite. For example, if we measure the vertical projection of the spin of one of the particles and we find it "up," then that of the other one *must* be down. Therefore, no matter how far apart the two particles are, we have effectively measured the vertical component of the spin of the second one without touching it, and it must react to a measurement of a horizontal spin component, say, just as if its vertical component had been measured. In the definition of EPR, the vertical spin component of the second particle has "an element of physical reality," for which the authors gave a minimal criterion: *"If, without in any way disturbing a system, we can predict with certainty . . . the value of a physical quantity, then there exists an element of physical reality correspond-*

ing to this physical quantity" (their emphasis). On the other hand, if we measure the east-west component of the spin on the first particle, by the same argument, the east-west component of the second acquires an element of physical reality. The quantum theory, however, does not allow us to determine these two quantities, the vertical and the east-west components of the spin, at the same time. Since they both have "elements of reality," the authors conclude that quantum mechanics cannot be a complete description of that reality. Alternatively we may ask, how does one account for the second particle's "knowledge" of what measurement has been performed on the distant first particle? Without instantaneous communication between the two particles (spooky action at a distance), we cannot account for it unless quantum mechanics is not the whole story.

The EPR argument also demonstrates, although the authors do not say so, that the intuitive idea of explaining the quantum mechanical effect of one measurement upon a subsequent one, as in the measurement of different components of the spin, by claiming that it is caused by the *disturbance* necessarily created in the process of measuring, is untenable. Unless, again, we postulate the instantaneous transmission of such disturbances over large distances, the measurement performed on one particle cannot very well have disturbed the other, possibly thousands of miles away.

Bell's Inequality

Thirty years after the EPR paper, the Irish physicist John Stewart Bell put the crucial distinction between classical and quantum mechanics on a firm mathematical basis. A simplified, schematic version of Bell's idea was devised by the American physicist N. David Mermin using the following thought experiment.

Our apparatus consists, to begin with, of an arbitrary emitter, which sends out two simultaneous signals or messages in opposite directions (see Figure 37). You may imagine these signals as two little balls, one sent to the left and one to the right, each with a message written on it, or you may envisage any other kind of transmitter of signals. The contents of the messages to be sent will be specified in a moment.

On each side of the emitter, at arbitrary but approximately equal distances from it, there is a receiver, that is, a box that receives the messages, balls, or whatever we use. Each receiver has a switch with three positions, and a red and a green light. The two receivers have

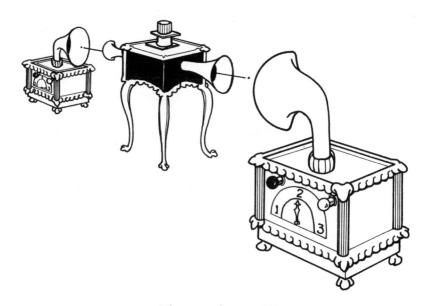

FIGURE 37. The setup for our EPR experiment:
a signal emitter and two receivers.

no connections of any kind to one another, no wires, no pipes, no radio transmitters, nothing. This implies that the message sent to an individual receiver could not tell it to respond in any way that would depend on the switch settings of the other receiver; neither receiver can know the settings of the other. The two can have their switches set in any of their three positions, completely independently of each other. In fact, we may want to introduce two independent random devices (such as roulette wheels or dice) that determine the settings of each of the two switches.

Here, now, is the order of events: (1) A pair of signals is sent simultaneously by the emitter toward the two receivers (one ball to each receiver). (2) The switches on the two receivers are set independently to any of their three possible positions before the signals arrive. (3) When a signal arrives at a receiver, either the red or the green light flashes on. (That the light flashes in response to the signal received can be ascertained, for example, by blocking one of the message paths and finding that no light goes on.)

This sequence of events is repeated many times, with all possible messages. The results are recorded by simply writing down, say "13(RG)," meaning "The switch on receiver #1 was in position 1, that of #2 was in position 3, the light on receiver #1 was red, and that on #2 was green." So, after many runs you have a long string of such symbols: 13(RG), 21(RR), 33(GG), 32(GR), ... When you examine them you find the following two characteristic facts: (1) Whenever the switches of the two receivers are in equal positions, the two lights flash equal colors. (2) If you disregard the positions of the switches and pay attention only to the colors of the lights, you find that half the time the lights of the two receivers flash the same color and half the time they flash different colors. You will have to take my word for it that according to quantum mechanics, signals employing particles with spin can indeed have the two mentioned properties.

The problem now is to account for these two observed characteristics by any "realistic" message without any communication between the receivers. Think, for example, in terms of two balls sent, with instructions written on them that tell the receivers which light to flash, depending on the setting of their switches. For example, the instruction RGR would mean: if the switch is in position 1, flash red; if it's in position 2, flash green; if it's in position 3, flash red. We may imagine other ways of sending the needed intructions.

Now let us look at these results and analyze them. Property 1 of the observed runs implies that the instruction in the two signals sent simultaneouly must be identical. Otherwise, for *some* equal switch settings the two receivers would flash different colors. Suppose then, for example, the instructions on both balls are RGR. There are nine possible combinations of switch settings; together with the resulting color flashes they are, in that case: 11(RR), 12(RG), 13(RR), 21(GR), 22(GG), 23(GR), 31(RR), 32(RG), 33(RR). You can see that there are 5 cases of equal colors and 4 of unequal ones. So for this particular instruction, RGR, the probability of equal colors is 5/9. It is easy to check that the same probability holds if the instructions are RRG, RGG, GRR, GGR, or GRG. If the instructions are RRR or GGG, which are the only other two possibilities out of a total of eight, then of course the probability of equal colors is 1. Therefore we find that in all possible cases of instruction sets that have the first of the observed statistical characteristics (that is, equal switch settings imply equal light flashes), the second one (that is, if the switch settings are ignored, then half the time the two lights flash the same color and half the time they flash opposite colors) is necessarily violated. In fact, equal lights must flash at least 5/9 of the time. We have to conclude that there can be no "independent reality" that accounts for the observations, unless we allow an instantaneous message to be sent from one receiver to the other, telling it what color it, in fact, flashed.

But that, too, violates Einstein's definition. It would constitute precisely a "spooky action at a distance." Note that the two receivers can, in principle, be very far apart.

In this example, on a classical, "realistic" basis equal colors have a probability that is at least equal to 5/9; this is a special instance of Bell's inequality. Since quantum mechanics allows a method of sending messages (as above) that leads to a probability of 1/2, it definitely violates Bell's inequality, and a number of experiments done in various laboratories over the last few years have confirmed this violation.

The implication, unrecognized in Einstein's time, is that one of two things has to be true, irrespective of further theoretical developments: either there is spooky action at a distance or else, at the microscopic level, there is no reality in the sense in which Einstein expected it. As Einstein put it in the EPR paper, "No reasonable definition of reality could be expected to permit this." Bohr, on the other hand, when asked whether the quantum mechanical algorithm could be considered as somehow mirroring an underlying quantum reality, declared,

There is no quantum world. There is only an abstract quantum mechanical description. It is wrong to think that the task of physics is to find out how Nature *is*. Physics concerns what we can say about Nature.

And Heisenberg took a similar position:

In the experiments about atomic events we have to do with things and facts, with phenomena that are just as real as any phenomena in daily life. But the atoms or the elementary particles are not as real; they form a world of potentialities or possibilities rather than one of things or facts.

The difference in world views between Einstein and Bohr has been characterized by John Wheeler using the following conversation among three baseball umpires, discussing ball and strike calls:

First: I calls 'em like I sees 'em.
Second: I calls 'em the way they *are.*
Third: They ain't *nothin'* til I calls 'em!

It is an essential feature of the quantum theory that it prevents us from ever forming a picture of the microscopic world that has all the ingredients we would normally require of "reality." This, more than

SPOOKY ACTION AT A DISTANCE

its destruction of the idea of a totally deterministic world, is the real philosophical revolution it has wrought.

In this chapter and the last we probed the two twentieth-century paradigm shifts in physics—relativity and the quantum—and discussed their profound consequences. Let us now turn to the effect that these two new theories have had on our ideas of what makes up the most basic constituents of the universe—elementary particles. The idea of their existence goes back to the ancient Greeks, but the picture we have of them now has been deeply altered by relativity and the quantum theory, as we will see in the next chapter.

8

WHAT IS AN
ELEMENTARY PARTICLE?

It all began with the Greek philosopher Democritus of the fifth century B.C. The ultimate constituents of the material world, he said, were hard, solid, invisible small particles that differed from one another only in shape and arrangement. They were indivisible (which, after all, is the meaning of their Greek name, *atomos*) and indestructible. Indeed the world, he thought, consisted of nothing but such atoms and the void. In this chapter, let us see what has happened to the concept of fundamental particles or indivisible corpuscles since the time of Democritus. We will find that not only have many new kinds of particles been discovered, different from any that classical Greek philosophy dreamed of, but the very notion of what we mean by the existence of such an entity as an elementary particle has been transformed.

Modern atomic theory, which systematizes the physical and chemical properties of matter on the basis of a few simple properties of the atoms, started some 2,200 years later, with the English chemist and physicist John Dalton, who explained the definite proportions with which elements react with one another chemically by postulating that all matter consisted of elementary atoms, which differed from one another by their weight. The atom, however, did not remain indivisible or indestructible very much longer. At the turn of the nineteenth century, the discovery of radioactivity by Antoine-Henri Becquerel, and of many of its properties by Marie and Pierre Curie, showed that atoms could be transmuted from one element to another; they were not the permanent building blocks imagined earlier.

At about the same time, J. J. Thomson found that cathode rays, emitted inside the now ubiquitous cathode-ray television picture tube and responsible for much of our entertainment, have a particulate nature. By deflecting them in a magnetic field, he recognized that they consisted of corpuscles with an electric charge that had a specific and constant ratio to their mass. The discrete nature of the electric charge was separately confirmed and measured fifteen years later by Robert Millikan, and we now call these negatively charged particles electrons. In Thomson's model of the atom, they were imbedded in a dough of positive charge. In 1911 Ernest Rutherford discovered, however, that the atom was mostly empty space, with almost all of its mass concentrated in its small nucleus, a new particle, thus destroying Thomson's raisin-pudding atom. In Bohr's subsequent model (already mentioned in Chapter 7), the atom consisted of a positively charged nuclear center surrounded by orbiting electrons. Together with some later refinements, such as Pauli's exclusion principle (as described in Chapter 7), it explained all of the properties of Mendeleev's periodic table of the elements, as well as the spectra of the light emitted by various atoms.

During the long interval of quiescense of the atomic theory of matter, dramatic changes had taken place in the concepts of the nature of light. Democritus had believed that light consisted of the cast-off skins of the atoms. The first modern systematic ideas concerning the nature of light were those of Christiaan Huygens, who described light rays in terms of waves, and Isaac Newton, who considered them made up of "small bodies." While Newton's speculations were far less fruitful than those of Huygens, his ideas were to sound a distant echo in the quantum theory initiated by Planck and Einstein 200 years later. In that theory, light has a dual nature, as a wave *and* as a "photon," a discrete quantum of energy whose measure is proportional to the frequency of its electromagnetic oscillation.

About sixty-five years ago this dual nature of light was joined with an equally dual nature of matter. Particles, too, were found to have a wave aspect, as first described by Louis de Broglie. Davisson and Germer discovered that electrons exhibit the same diffraction phenomenon that had originally led to the victory of Huygen's wave theory of light over Newton's "small bodies." Except for the fact that the mass of a photon is zero, there was now little difference between it and the corpuscles of matter.

The discovery of new particles then began to accelerate. In 1931 Wolfgang Pauli postulated the neutrino in order to account for a systematic loss of energy and angular momentum in the radioactive beta-decay of nuclei. The positron, predicted by Dirac as the antiparticle of the electron (though many theorists originally thought this antiparticle might be the proton, whose charge has exactly the same magnitude as the electron's but the opposite sign), was found by Carl Anderson in 1932, and during the same year James Chadwick discovered the neutron. Then came the first mesons: in 1937, the mu-meson (now called muon), which was initially misunderstood to be the carrier of the nuclear force as predicted by Hideki Yukawa, and ten years later the pi-meson (now called pion), which was the real McCoy of Yukawa.

The existence of the antineutron and antiproton, both anticipated on the basis of Dirac's theory and confidently expected by most physicists, was experimentally confirmed in 1956. By that time the first few of a whole series of new "strange" particles had been found: the lambda and sigma "hyperons" and the K-mesons. The list of the "hadrons," or strongly interacting particles, has by now grown much longer than that of the atoms. Later arrivals were called J/ψ (psi), Σ (sigma), Y (upsilon), and other letters in the Greek alphabet; sometimes their discovery caused enough excitement to make the front page of the *New York Times*.

With respect to the large and growing list of experimentally detected particles, we are now at a stage that is comparable to the discovery of the periodic table of the elements. The particles have been classified and ordered and their properties have been systematized by schemes that are often mathematically very sophisticated. Sometimes these schemes even predict the existence of new particles and their properties, later confirmed in experiments, just as Mendeleev predicted elements that were then unknown. Murray Gell-Mann received the Nobel Prize after successfully predicting the omega-minus particle and some of its properties. However, our understanding of the origin or the physical basis of the most useful classification schemes devised is far from complete. There is now thought to be an analogue of Bohr's atomic model at the bottom of the bewildering variety of particles found—the so-called *standard*

164

model of elementary particles, which is based on the field-theoretical ideas discussed near the end of Chapter 4.

How Are Particles Usually Detected?

Since the particles we are talking about are much too minute to be seen even under the most powerful microscope, what do we mean when we say they exist? The answer is not as intuitive as you might imagine.

J. J. Thomson inferred the corpuscular nature of cathode rays because they were bent by magnets in a manner that indicated whatever they were made of had a definite ratio of electric charge to mass. Millikan saw the discreteness of the electric charge directly in the behavior of his oil drops in an electric field, each with only one or a few electrons on it. What is meant by the existence of an electron is clearly the fact that the electric charge comes in distinct, finite chunks, always of the same size. Associated with these quanta of negative charge there is always the same discrete amount of mass and angular momentum, or spin, which was discovered fifteen years later. They are always exactly the same. Furthermore, electrons are stable; they will not, by themselves, fall apart or disintegrate. As they pass through a vapor-filled cloud chamber or a liquid-filled bubble chamber (or a glass of beer, which gave rise to the invention of the bubble chamber), their charge will cause sufficient ionization of the molecules they pass through to produce visible droplets or bubbles; they leave tracks. In a magnetic field that causes the paths of charged particles to curve, the tracks are visibly bent (see Figure 38). Who could doubt the existence of such particles?

The case of the neutrino was more problematic. Certain experiments on radioactivity showed that in the beta-decay of nuclei an electron was emitted, but some energy and some angular momentum were missing. Pauli interpreted the details of the evidence by postulating the existence of a massless and electrically neutral particle that was emitted at the same time as the electron. Since this particle has no charge, it would never leave a track in a cloud chamber. Its existence was gradually accepted, although for many years there was no more direct evidence than the breakdown of certain conservation laws if it were nonexistent. Understandably, some skepticism lingered in the minds of a few physicists until the experimental detection, twenty-four years later, of inverse beta-decay. In this process, the neutrinos produced in some earlier nuclear beta-decay collide with

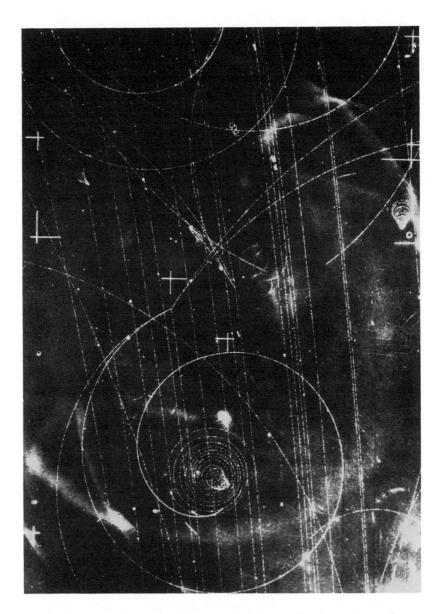

FIGURE 38. Photograph of the traces of electrically charged particles in a hydrogen bubble chamber in a magnetic field. The almost-straight tracks are made by the fast particles of the beam and the curving tracks, bent by the magnetic field, are those of slower electrons and pions produced in collisions.

WHAT IS AN ELEMENTARY PARTICLE?

nuclei in the laboratory to produce the opposite reaction, in which a positron (the positively charged twin of the electron mentioned earlier) is produced and some energy and angular momentum are added. After that the "reality" of the neutrino could no longer be doubted. By now, millions of events induced by neutrinos from the sun or from a supernova have been studied in laboratories all over the world.

Traditionally, the principal means of detecting electrically charged particles is a track in a cloud chamber, a bubble chamber, a spark chamber, or a photographic emulsion (see Figures 38 and 39). A magnetic field, combined with microscopic observations of the tracks, permanently visible in a photograph, allows the measurement of the mass and the electric charge of the invisible particle that is the cause of the track. Other attributes that can be measured are its spin (see Chapter 7), its magnetic moment (which means that it behaves like a small magnet), and other so-called quantum numbers that codify its behavior with respect to certain physical laws.

More recent methods of detection use counters that are directly connected to computers, with no visible tracks at all, except perhaps in the minds of the physicist and the computer. Figure 40 shows the first four events in the detection of the Z^0 (pronounced "z-zero") particle, which is responsible for "neutral currents" in weak interactions. The Z^0s are here produced in proton–antiproton collisions, and since they are electrically neutral, they cannot be directly detected. Instead, their presence is signaled by their subsequent decay into electron–positron pairs. The four pictures show plots of the deposit of energy in "bins" or detector cells in various positions, with sharp spikes found in those bins that, by the conservation of energy and momentum (to be discussed further in Chapter 10), correspond to the decay of four particles of the same mass (within experimental error) into an electron and positron each, and no additional particles. For this discovery at the CERN laboratory in Geneva, Switzerland, Carlo Rubbia and Simon van der Meer shared the Nobel Prize of 1984.

No Permanence, No Individuality

The quantum numbers of newly discovered particles are usually inferred from their scatterings and the distribution of their production, as well as of their decay products. Thus, one makes use of what is perhaps the particles' most fundamental general property, discovered and theoretically understood about 60 years ago, namely that all of them can be created and destroyed, both by nature and in the labo-

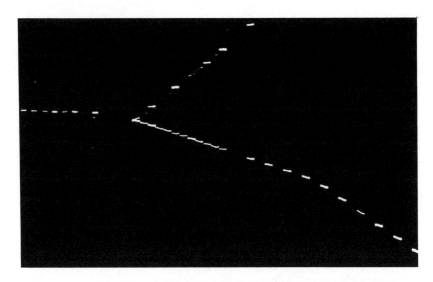

FIGURE 39. The tracks of charged particles in a magnetic field in a spark chamber. The large gap is the invisible path of a neutral particle moving to the right and decaying into two oppositely charged particles.

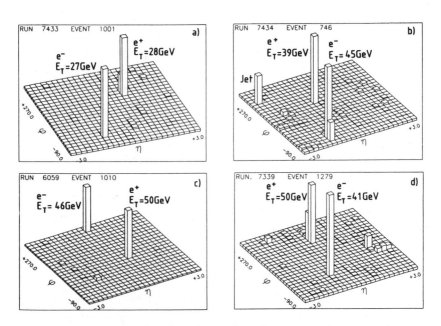

FIGURE 40. The first four detected Z°-decays into electron-positron pairs. See the text for the meaning of these four pictures.

WHAT IS AN ELEMENTARY PARTICLE?

ratory. Subject to certain general conservation laws, such as that of momentum, energy, and electric charge, even electrons may be created. As we have already discussed in Chapter 4, under certain conditions, an energetic photon may create a pair of an electron and a positron. Conversely, an electron may be destroyed in a collision with a positron and give rise to a photon or, if energetic enough, even to a pair of a proton and an antiproton. The most dramatic evidence of the creation of particles is a "shower," in which an extremely energetic cosmic-ray particle (usually a proton) coming from outer space hits a nucleus in the terrestrial atmosphere, producing hundreds of particles of all kinds in the original and subsequent collisions. Their simultaneous arrival, scattered over a large area on the earth's surface, can be detected in counters. So, it turns out that we can make particles as well as nature, which produces them all the time. Do these particles have any kind of individuality?

It is a basic tenet of the quantum theory—and its consequences are well borne out by the experimental evidence—that particles of the same kind, such as electrons, are not only similar to each other but indistinguishable one from another. A proton cannot be tagged and identified as coming from an individual source. Indeed, an electron seen earlier and apparently again later may in the meantime have done weird things, such as emitting a photon that may, in turn, have created an electron–positron pair, the positron of which may have been annihilated together with the original electron, so that the electron seen later may "really" be the one created as part of the pair. The word "really" here has no meaning. There is, as a matter of fundamental principle, no way of telling which electron is which.

Thus, we have lost two of the most significant standard attributes of particles: permanence and individual identity. This, however, is only the beginning of the story. Electrons, if they are not permanent, are at least stable. That is to say, if left alone, they will not disintegrate but remain intact. The same is generally thought to be true of the proton, though there is now some doubt about the proton's stability, and it is undergoing searching and difficult observation. Fortunately, if protons turn out to be unstable, their average lifetime is very much longer than the age of the universe; so we need not worry about suddenly disappearing in a puff of particle-smoke. Electrons, and probably protons, are among the very few stable particles. Others that are thought to be stable are the positron, the antiproton (or perhaps not), the neutrino, and the photon. A neutron that is not in an atomic nucleus will disintegrate into a proton, an electron, and an antineutrino, with a life expectancy of about 13 minutes. This is the analogue of the radioactive decay of some atomic nuclei. Most of the remaining particles have much shorter lifetimes. The muon decays, on the average, after about 10^{-6} seconds, the lambda after about 10^{-10} seconds, and some of the other hyperons after some 10^{-23} seconds.*

How is it possible to detect the presence of particles that exist for as short a time as 10^{-23} seconds? Thereby hangs a tale that requires some explanation and an excursion into elementary physics. We will have to understand what a *resonance* is, because that is the characteristic sign of the presence of such a particle.

Resonances in a Pendulum

Think of a simple pendulum, a weight hanging from a thread. Like a clock, it has a specific frequency of oscillation, say f. Suppose we take a small hammer and gently begin to hit the resting weight rhythmically, with another frequency—g. Since the hammer will hit the pendulum bob the second time when the latter is in the middle of its motion, and then again when it is somewhere on its way, not much of anything will happen on the average, aside from somewhat irregular swinging. Suppose we try this with all sorts of different frequencies g, and we plot the energy that the pendulum absorbs from the hammer as a function of the frequency g. The curve will be fairly flat, except when you get to the frequency f. Then the first hammer stroke will set the pendulum in motion and the second stroke will come exactly when the pendulum is ready to swing forward again in any case, thus reinforcing the first, and so on, like pushing a child on a swing. The amplitude of the pendulum will get larger and larger. This means that a lot of energy is being transferred from the hammer to the pendulum, that is to say, the pendulum absorbs a large amount of energy from the hammer at that frequency $g = f$, its "proper" or "characteristic" frequency. One says that the pendulum, or any vibrating system, has a *resonance* at f. It is the same effect that can make your car shake very badly at certain vibration frequencies caused by the road, and that may make a bridge disintegrate if

---

*In order to get some feeling for these periods of time, it is useful to compare them to timescales like the following: the present age of the universe is about 10^{18} seconds; human life expectancy is about 2×10^9 seconds; the length of a heartbeat, about 1 second; the passing of a single light wave takes about 10^{-15} seconds.

WHAT IS AN ELEMENTARY PARTICLE?

specific vibration frequencies are excited (as in the Tacoma-Narrows bridge disaster in 1940).

If there were no friction, this would go on forever, that is, all the energy would be absorbed and the resonance frequency would be absolutely exact, sharp. However, there is always some friction or damping of the pendulum. Left alone, it will not swing forever but slowly lose energy to its environment and come to rest. This has the consequence that its proper frequency is not quite sharp. A curve like Figure 41, which depicts the oscillatory motion of a damped pendulum, can be decomposed into a superposition of functions, each with one exact frequency and constant amplitude. The frequencies "contained" in this slowly damped curve are all very close to f. As a result, the motion has a small range of proper frequencies around f, not just one sharp f. If we plot the amplitude or strength of each frequency contained in the pendulum oscillation as a function of this frequency, it will look something like Figure 42: a sharply peaked curve with a width of the peak (at half of its maximum) equal to Γ.

Let us look back at the energy loss of the pendulum, as drawn in Figure 41. The time T at which it has lost half its energy is called its *half-life*. One can demonstrate mathematically that there is a simple relation between the width of the peak of frequencies contained in the oscillations of the pendulum and its half-life: $2\pi T\Gamma \simeq 1$. (The twiddle means "approximately equal to.") In other words, the width Γ of the resonance is inversely proportional to the lifetime: $\Gamma \simeq 1/(2\pi T)$. If there were no friction or damping, the half-life would be infinite, that is, the oscillations would never decrease in amplitude; the resonance width Γ would then be zero. There would be only one sharp characteristic frequency. On the other hand, a short lifetime means a large width.

If we have a damped oscillator, which has a range of proper frequencies, it will absorb energy from the hammer that sets it in motion not just at the one frequency f but in a frequency interval around f of width Γ. The absorption, plotted as a function of frequency, has a sharp resonance peak that looks like Figure 42.

Resonances in Atoms and Molecules

The principle we have just seen exemplified by a simple pendulum is exactly the same for more complicated oscillating systems. They and the "hammer" that sets them in motion need not be mechanical but may be electromagnetic, for example. Atoms have characteristic fre-

Amplitude

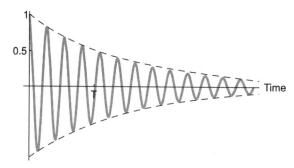

FIGURE 41. A typical plot of the motion of a damped oscillator: its position as a function of the time. After the time $t = T$ its amplitude of oscillation has decreased to half of its starting value; when $t = 7T$, it has dropped to less than 1%.

Amplitude

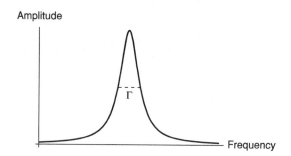

FIGURE 42. Schematic plot of the strengths of the various frequency constituents of the damped oscillations depicted in Figure 41. The width Γ (at half maximum) of the curve is related to the decay-time T by $\Gamma = 1/(2\pi T)$.

quencies, and the plots of the absorption of light by atoms look just like the ones described above. These plots are the absorption lines for light by atoms or molecules (see Figure 43). A given kind of atom may have more than one such resonance frequency.

What is the analogue of the friction that caused the damping of our pendulum in the case of atoms? The oscillating system in an atom is the collection of electrons that move around the nucleus. Because electrons are electrically charged, an electromagnetic wave such as light passing over an atom can excite it and make the electrons oscillate more strongly (somewhat figuratively speaking). But since

WHAT IS AN ELEMENTARY PARTICLE?

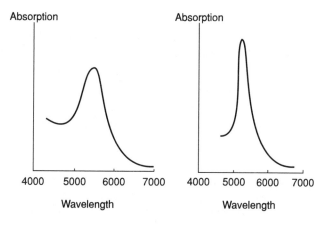

FIGURE 43. These curves show the absorption of light (left) by a colloidal gold solution and (right) by ruby glass.

they are electrically charged, the electrons will also emit electromagnetic radiation and lose energy, which causes damping. The mechanism for the damping is the emission of radiation, that is, light, via the same coupling of the electrons to the electromagnetic field that makes their initial excitation possible, namely their electric charge.

The magnitude of this charge, and thus the damping, is very small, so that the half-life (by which, according to quantum mechanics, we now have to mean the time at which half of the excited atoms have become de-excited) of the excited system is very small and the absorption lines, as well as the emission lines, are very narrow. Atoms absorb and emit light at characteristic colors that are almost exact, and for many purposes they can be considered exactly sharp. Since the half-life of an excited atomic state is of the order of 10^{-8} seconds, the line width is about 10^7 Hz,* which is equivalent to a relative uncertainty in the emission frequency of about one part in 10^8.

At this point we must add to the picture two fundamental facts. The first, coming from quantum mechanics, is that the energy E of simple oscillating systems comes in packages or quanta of fixed size proportional to the frequency f, $E = hf$, the constant h of proportionality being Planck's constant. Such packages of electromagnetic energy are the photons already mentioned earlier. For mechanical vibrations in solids, the quanta are called *phonons*, yet another set of particles.

*Recall that Hz, *hertz*, is the unit of frequency, one oscillation per second.

WHAT IS AN ELEMENTARY PARTICLE?

The second fundamental fact to be added, from the theory of relativity, is the famous relation, also mentioned earlier, between energy E, mass m, and the velocity of light, c, namely, $E = mc^2$. An object of mass m, even when at rest and subject to no external forces, has a "rest energy" equal to mc^2. (For photons, which always travel at the speed of light c, it turns out that one must set $m = 0$. Particles of nonzero rest mass can never attain the speed of light; photons, on the other hand, can never come to rest.)

Resonances in Scattering

We now have all the conceptional ingredients necessary: the idea of a resonance in a vibrating system, the quantization of energy and its relation to the frequency, and the relation between the rest mass of a particle and its energy.

Instead of shining light on atoms, imagine bombarding them with electrons. In that case, too, we can excite the atoms at or near their characteristic resonance frequencies, which we now understand to mean *resonance energies*. So if we vary the energy of the bombarding electrons, they will lose energy at certain resonance lines. The atoms will absorb that energy, and after a characteristic half-life, half the excited atoms will have lost it again by emission of photons of the appropriate energy. What has happened is called *inelastic scattering* of the electrons; they have lost a characteristic amount of their energy, and that energy has been converted, after a short delay, into photons, that is, light. It's as though a group of children, each with different-colored ice cream cones, were to mingle with another group of larger children who hanker after strawberry ice cream; after a while, all the pink ice cream cones of the first group will have been licked by others, and the happy recipients would be walking around whistling. Ice cream of a certain range of colors has been converted into musical sounds of certain melodies!

Another process may also occur. At certain characteristic energies, an electron can be captured by the atom and form a negatively charged ion. After a short time, the electron will be able to escape again and go on its way, marching to its own drummer; if it could get into the atom, it must also be able to get out. (This is the analogue of the absorption and subsequent re-emission of a photon by the atom.) The fact that the electron was captured and spent some time in the atom leads to a large deflection from its original path, which can be observed as a large amount of scattering. If we bombard

WHAT IS AN ELEMENTARY PARTICLE?

atoms with a beam of electrons, we can measure how many have been scattered. At certain characteristic energies (that is, frequencies) there will be much more scattering than at others. If we plot the amount of scattering against the energy of the incident electrons, we will observe resonances, just as we did before. The energy of the resonance line gives us the energy, and hence the mass, of the ion that was temporarily formed. The width Γ_E of the resonance line allows us to infer the half-life of the ion. Using $E = hf$ and $T\Gamma \sim 1/(2\pi)$, we obtain $T\Gamma_E \sim h/2\pi$. (This is also known as one of Heisenberg's *indeterminacy relations*.) In energy as in frequency, the longer the half-life of the unstable system formed, the sharper the resonance.

We can now understand that by observing resonances in the scattering of particles, we learn of the existence of systems of these particles that can remain together as one unit for some length of time. Physicists make such observations by accelerating electrons, or protons, or other ions, in machines–such as a cyclotron or a synchrotron, or a linear accelerator—and shoot them at each other at varying energies. The measure of the amount of scattering observed is called a *scattering cross section*: it represents the cross-sectional area these particles appear to present to one another in their collisions at various energies. If we plot such a cross section as a function of the energy, a sharp bump may appear. This will generally be interpreted as evidence for a resonance, and hence of an entity that will stay together with a half-life equal to Planck's constant divided by the width of the bump.

There are, of course, reasons other than resonances for variations of a cross section with energy. When are we justified in concluding that an observed bump in a scattering curve represents a resonance? There is an effect, already mentioned parenthetically, that is associated with a real physical resonance, namely, a delay in the time of emergence of the scattered particles because they are caught in an almost stable system. This could, in principle, be measured either directly or via some indirect side effects. However, such measurements are much more difficult to perform than those of a scattering cross section. If we know something about the range of the forces involved, it is possible to draw more immediate conclusions.

Let us assume that the range of the forces is D. Then the normal time it would take two high-energy particles, traveling near the speed of light c, to traverse one another's force-fields is approximately D/c. If there is a resonance, the particles spend much more time together than D/c before flying apart; this delay is approximately the same as the half-life of the system. Hence, if the width Γ_E of a sufficiently high peak in a cross-section curve is small compared with hc/D, it can be assumed to be a genuine resonance. As a practical matter, a noticeably sharp peak in an otherwise gently varying cross-section curve is usually interpreted as a resonance.

So here is the explanation of how an entity that "lives" for a certain length of time can be discovered in a scattering experiment: it is observed as a resonance line (that is, peak) in a plot of the cross section. Let's look at some numbers in their rough order of magnitude. Nuclear energies are measured in keV (1,000 electron Volts) or MeV (million electron Volts). A half-life of several years of a radioactive nucleus corresponds to a line width of 10^{-23} eV; this is much too sharp to be resolvable by any feasible measuring device. A half-life of 10^{-10} seconds still corresponds to a width of 10^{-5} eV, and that also is too sharp to be detectable as a resonance. However, if such particles travel at a speed approaching that of light (as they do in practical experiments), they will traverse distances measured in millimeters or more before they disintegrate. Hence, if they are electrically charged, as most are, their tracks can be made visible in a cloud chamber, a bubble chamber, or a spark chamber. We can "see" them directly, so to speak, as we have earlier discussed. If they are electrically neutral, their "paths" can be inferred from the correlation between the point of their production by visible charged particles and the point of their decay into charged products. If they exist for times that are too short to give rise to visible tracks, say less than 10^{-13} seconds, but not short enough to lead to a resolvable and visible resonance line, they are hard to detect. However, when their half-life is less than about 10^{-18} seconds, they will lead to recognizable resonances; even though the experimental energy resolution may not be as fine as the resonance width, the narrow, high peak will be recognizable in the form of a lower, broader bump on an experimental curve.

Again, the theory of relativity enters with an interesting effect. As we discussed in Chapter 6, it predicts that a moving "clock" slows down; the faster it moves, the slower it seems to go. Consequently, if a particle of very short half-life moves rapidly, its decay, as seen in the laboratory, is retarded and its half-life is extended. (For particles, life in the fast lane is longer rather than shorter!) This phenomenon, already mentioned in Chapter 6, is experimentally well established. It follows that if an electrically charged particle of very short life is produced at high enough speed it will, in principle, always be able to

WHAT IS AN ELEMENTARY PARTICLE?

traverse a distance of visible length before decaying. However, as a practical matter, the energy necessary to give it such speed is often too high.

Voilà: A Particle

The objects we are now describing may not be what Democritus had in mind when he talked of particles, but they have a (directly or indirectly) measurable energy, and hence both mass (which is uncertain by the amount of the resonance width) and a measurable half-life. They also have a number of other properties that can all be inferred from an analysis of the details of the scattering experiment: electric charge, magnetic moment, angular momentum, and other quantum numbers. It is the collection of these properties that a physicist today uses to identify a "particle." We can associate several different kinds of "sizes" with such an object. The first is the length $h/2\pi mc$, if its mass is m, which is known as its *Compton wavelength*; for the electron it is 3.86×10^{-11} cm. (Arthur Compton had discovered in 1922 that electrons scatter photons, as they do other particles, thereby changing their energy and, hence, frequency. For this fundamental discovery, which is a powerful confirmation of the particle-nature of light, he received the Nobel Prize in 1927.) The second "size" is the scattering cross section, which is an area. It is not constant but depends on the collision partner and varies with energy; it is measured in barns (10^{-24} cm^2) or micro-barns (10^{-30} cm^2) or nano-barns (10^{-33} cm^2). From an analysis of the scattering cross section of an electrically charged particle, such as a proton, it is possible to infer the extent to which the charge of the particle is either concentrated in a point or distributed over a finite volume; in that way it may be said to have a "charge radius," a third kind of "size." Electrons are, so far as is known, pure point particles in this sense: their charge is concentrated at a point; protons, on the other hand, have their electric charge distributed over a ball whose radius is approximately 10^{-13} cm. They have no shape (other than spherical) nor individuality. All that is left of the classical idea of a particle is the specificity and permanence of its properties. As a dust particle is distinguished from a whiff of smoke by the sharpness and permanence of its contours, so an electron differs from an electrostatic field by its well-defined mass, charge, and spin. However, even the mass of a neutron, which, when left all by itself outside an atomic nucleus has a half-life of 13 minutes, is "smeared out" by about one part in 10^{27}. (This intrinsic uncertainty of its mass is very small compared with the precision with which that mass can be measured experimentally.)

If there are two particles of almost identical properties, a very peculiar identity crisis may arise that can be explained only quantum mechanically. A case in point is that of the neutral mesons called K^0 and $\overline{K^0}$, which are born in the decay or in the collision of other particles. Each of them is a quantum-theoretical "superposition," half and half, of two other mesons called K_s^0 and K_l^0, which decay with very different half-lives. The half-life of the K_s^0 is 10^{-10} seconds, and that of the K_l^0 is 500 times longer. Hence, after some 5×10^{-10} seconds, or 15 cm of travel at close to the speed of light, the K_s^0 will have almost entirely disappeared and only the K_l^0 will be left. But the K_l^0 particles, in turn, are a superposition, half and half, of K^0 and $\overline{K^0}$. Therefore, after 5×10^{-10} seconds, or 15 cm of travel, particles born as K^0 become half K^0 and half $\overline{K^0}$. The K^0 and $\overline{K^0}$ interact differently with other particles, and the change in identity can be detected. Such objects have a chamleonlike character that makes the name "particle" intuitively somewhat confusing.

Which of the Particles Are Elementary?

Because of the many particles that have been discovered during the last forty years, the question has naturally arisen: which of them if any, are really "elementary"? We used to consider atoms as indivisible and elementary; that was the main point of the atomistic idea and the origin of their name. Then atoms were discovered to consist of a nucleus and electrons. Later the nucleus was discovered to consist of protons and neutrons. For about twenty-five years after that, protons, neutrons, and electrons, and possibly some mesons, were thought to be the elementary particles. But then the number of newly discovered elementary particles proliferated to such an extent that it was natural to conclude either some small number of as yet undiscovered particles must be truly elementary, constituting those seen in the laboratory, or else the whole idea of elementary particles made no sense. The concept that all the particles are really on the same footing, that none is, in any fundamental sense, more "elementary" than others, is sometimes referred to as "nuclear democracy." This question is still to some extent open, but there is the following intriguing aspect to it.

As we have seen, almost all the known particles are unstable. That is to say, after a certain length of time they spontaneously decay into others. Of the particles we have discussed, only the electron, positron, neutrino, photon, and probably the proton and antiproton are absolutely stable. It seems impossible to invent a theory in which stable particles, and only stable particles, are "elementary." In such a theory the neutron would be ruled out automatically, whereas the proton (if it turns out to be stable) could be a candidate; to make a fundamental distinction of this kind between two such closely related particles would be bizarre. But if, on the other hand, an unstable particle is to be elementary, one must surely demand that it have no "memory." An internal mechanism that seems necessary for a memory would make it nonelementary. But if there is no memory, the particle does not "remember" when it was born, and hence the probability of its decay must be constant in time. It follows mathematically that the law of its decay as a function of time must necessarily be an exact exponential, the reciprocal of the natural growth function. Now all the experimentally found decay laws of unstable particles are in fact exponential laws to a very excellent approximation, so good indeed, that no deviations have ever been detected. But quantum mechanics also tells us that the decay law cannot be an exact exponential; at least no way has yet been discovered by which it could possibly be exact. That seems to rule out the idea of an unstable elementary particle, unless the contemporary framework of the quantum theory has a basic flaw.

Many physicists are now loath to use the phrase "elementary particle"; they prefer the term "fundamental" for some of the many particles discovered during the last 40 years. It would, of course, be as implausible to call all of them "fundamental" as to use that name for all the atoms of the chemical elements; it would seem much more likely that there are other, more fundamental particles constituting those which have been found. Either that, physicists reasoned, or there is "nuclear democracy," and no particles are more fundamental than others.

Present theory indeed postulates the existence of two basic kinds of particles, "quarks" and "leptons," each of which comes in six different "flavors." (The name *quark* was chosen by Gell-Mann from a line in Joyce's *Finnegan's Wake;* the word *lepton* originally denoted a small Greek coin.) The quarks are the building blocks for the construction of models and the basis for classification schemes of all the many known particles, just as electrons and nuclei are for the construction of the periodic table of the elements. However, while leptons have been detected in experiments (they include electrons and neutrinos), no isolated quarks have yet been found and no one has succeeded in forming quark beams. Physicists have searched for them in vain. The presently accepted explanation for this is that, in contrast with electric, gravitational, and other known forces, the pull between quarks does not get weaker with increasing distance but remains essentially constant. As a consequence, it takes an ever-increasing amount of energy to move those that are caught in one another's embrace farther and farther away from one another, and they cannot get out of the resulting "bags" that hold them inside the particles we detect. Furthermore, once you have invested a sufficiently large amount of energy to pull two quarks apart, that energy is converted into the creation of a quark–antiquark pair. The quark of that pair may stay with one of the original two quarks, and the antiquark may join the other. Whereas your aim was to get two isolated quarks, you now have a quark–antiquark pair in one place and two quarks in another, but still close together.

Bosons and Fermions

But what about such particles as the photon, the quantum of light? Are these not also elementary particles? According to quantum-theoretical thinking, the wave–particle duality applies both to the entities that used to be regarded as honest particles, like electrons, and those that used to be thought of as waves, like light. This duality indeed holds for both, and yet there is a fundamental difference between the two. Matter is made up of some of the particles described above, electrons and quarks; these belong to a category called *fermions*, after the Italian physicist Enrico Fermi. Such particles obey Pauli's exclusion principle, which prevents any two of them from being in precisely the same physical state. One of the results of this is that two of the same kind cannot occupy the same place, and matter cannot collapse completely.

The particles in the other category are called *bosons*, after the Indian physicist Satyendranath Bose. These are not subject to the exclusion principle, and they are the transmitters of the various kinds of forces that particles exert upon each other: photons are the bosons of the electromagnetic force-field, pions of the strong nuclear force, gravitons of the gravitational force, gluons of the interquark forces, and so on. What complicates this situation is the fact that these bosons also exert forces on one another, some of them directly and some of them indirectly, by creating fermion pairs. (The previously mentioned Delbrück scattering of light by light is an example of the indirect interaction.)

According to present-day thinking, then, the particles that may be called "fundamental" are the leptons and quarks (which are fermions), the photon, the gluon, the graviton, and a few particles (including the Z^0 mentioned earlier) associated with the weak interaction (all of which are bosons). The pion, which used to be regarded as fundamental, is not among them; it is made up of a quark and an antiquark.

The Particle Concept Has Heen Transformed

What has happened to the particle concept? Surely, Democritus would not recognize it. Earlier in this chapter I mentioned the phonon, a quantum of sound (or of mechanical vibration in a solid), just as the photon is a quantum of light (or of electromagnetic vibration), and just as a pion is a quantum of the nuclear force-field. Is there any reason to consider phonons any less "real" as particles than photons or pions? Of course, they cannot get out of the solid and cannot be detected in splendid isolation, but neither, it appears, can the quarks get out of a proton. In the minds of skeptics this may raise doubts about the meaning of the assertion that such entities "exist." After all, deep-down many of us are "naive realists" who have a hard time believing in the existence of objects we cannot lay our hands on, at least in a manner of speaking. However, physicists are unconcerned with such metaphysical concepts as *existence*. The building blocks that form our basic concepts of the world are selected according to whether they work and form a coherent network of ideas. Physicists are not in the business of discovering "ultimate reality," whatever that may mean. Our ideas of reality necessarily have to be formed on the basis of what we know, and physics is the science that probes nature most deeply. If, in our search for deeper structure by means of more and more refined experimental tools, the concept of what we call a particle becomes unintuitive and somewhat abstract, that is presumably how it has to be. We must not insist that our preconceived notions, based as they are on very much less refined sensory experiences, should be sufficient to encompass all natural phenomena. Democritus may have been right, at least in part, but the meaning of what he asserted has to be reinterpreted in accordance with what we now know.

In this chapter we examined in some detail to what extent the

181

ancient idea of elementary particles has been transformed by the quantum theory in conjunction with the theory of relativity. The astoundingly large number of different such particles, we saw, made it necessary to descend to a deeper level of analysis and to assume the existence of more fundamental entities, by means of which the particles that had been discovered could be classified analogously to the periodic table of the elements constructed by means of Bohr's model of the atom made up of electrons and a nucleus. That leaves us with the unanswered question of the basis on which the existence of these more fundamental "subparticles," which were themselves undetectable by direct means, could be postulated. We shall return to that question in the final chapter. First, however, let us turn to the exploration of some physical phenomena whose explanation requires ideas that we have so far avoided: we must study the behavior of systems whose parts act *collectively*. This will open up a whole new vista on our concepts of the world.

WHAT IS AN ELEMENTARY PARTICLE?

9

COLLECTIVE PHENOMENA

It is sometimes said that science, by its very nature, is *reductive,* which, in the view of those who accuse scientists of the sin of reductionism, is a grave shortcoming. There is, of course, some truth to this allegation, but there is no need for the practitioners of science to be apologetic about reducing complicated phenomena to their simpler components, wherever possible, in order to understand them better. Indeed, the opposite approach, in which every multifaceted effect is regarded as totally new and not reducible to parts already understood or easier to comprehend is neither desirable nor fruitful.

Nevertheless, there are certain natural phenomena that are resistant to such reduction and about which it can truly be said that the whole is more than the sum of its parts. While the understanding of such physical effects has taken and is still taking much longer than that of those which can be reduced to the sum of their parts, attempts at such understanding have been pursued for a hundred years and are at the forefront of much current research in physics. Some of these we shall discuss in this chapter.

Almost all progress in our comprehension of collective or cooperative effects in nature is based on the quantum theory, which is why such enlightenment had to wait until the twentieth century. Because these effects, by their very essence, generally involve a multitude of particles, it is not surprising that they occur mostly in the part of physics dealing with aggregates, the discipline usually called *condensed-matter physics*. They are not, however, completely confined to that field. Another area in which collective phenomena play important roles is nuclear physics, from which our first examples will be taken.

Neutrons in Atomic Nuclei

The neutron, whose discovery by James Chadwick in 1932 ushered in the age of nuclear physics, is not a stable particle in isolation but decays after an average lifetime of 13 minutes into a proton (slightly lighter than the neutron and carrying positive electric charge), an electron (very much lighter by a factor of about 2,000 and carrying a negative electric charge of equal magnitude), and an antineutrino (massless, or almost so, and electrically neutral). Why do the neutrons that appear in every atomic nucleus (except that of hydrogen, whose nucleus consists only of a proton) not decay and leave as a final residue of all matter only hydrogen and antineutrinos? In some nuclei, of course, neutrons do decay, making the material consisting of these atoms radioactive. But most materials are not radioactive, and stars do not convert to hydrogen after 13 minutes in a puff of antineutrinos. How can we account for this?

The explanation of this apparent puzzle may be regarded as the simplest collective effect in physics. In each atomic nucleus the neutrons and protons exist in certain discrete states, just as the electrons do in the outer part of the atom. According to Pauli's exclusion principle, each state can be occupied by only one particle (if we include its spin as part of the specification of a state). Because of the conservation of energy and the equivalence of energy and mass by Einstein's $E = mc^2$, when a neutron decays, the highest energy that the emitted proton can have is equal to that of the decaying neutron minus the "rest energy" of the emitted electron. Therefore, if all the energy levels of protons in the nucleus up to that maximal value are already occupied, there is no room for an emitted proton and the neutron simply cannot decay! The city of the nucleus has strict population limits; when they are reached, births are no longer allowed and nature enforces its birth-control laws with the utmost rigor. Thus, the neutron in the cage of a nucleus does not behave as it does "in the wild." Pauli's exclusion principle, we find, not only provides a cornerstone for all of chemistry, via the periodic table of the elements, but is also responsible, in part, for the stability of atomic nuclei.

There are other, more complicated collective effects in nuclear physics. Whereas in a heavy atom we may, as a first approximation, regard each electron as moving in the strong electric field produced by the relatively large electric charge of the small nucleus, the protons and neutrons inside the nucleus are not attracted by a force from the center but move and stay together under the mutual attraction provided by the strong nuclear force they all exert upon one another. The resulting "many-body problem," though relatively easy to formulate quantum mechanically in the abstract, is extremely difficult to solve, and nuclear physicists have had to divise models in which cooperative effects of the particle population of the entire nucleus are taken into account. (With the advent of quarks as the constituents of neutrons and protons, this picture has become even more complicated.) One particular collective influence changes the "effective" mass of a particle inside the nucleus so that it is not the same as outside when the particle is in isolation. It is as though your weight in crowded New York were different from your weight romping on an empty beach.

Whereas the first model, proposed by Bohr, pictured the nucleus as analogous to a drop of liquid in which the protons and neutrons moved more or less randomly inside a balloonlike bag, in 1949 Maria Goeppert-Mayer introduced the very fruitful *shell model,* for which she won the Nobel Prize in 1963. It was based on the idea of replacing the many individual forces that the particles exert upon each other by an average overall force-field that each neutron and each proton feels. Quantum mechanics then automatically allows for each particle only discrete levels, arranged in shells like the layers of an onion, analogous to the allowed levels of the electrons in an atom. This picture explained the existence of certain "magic numbers," nuclei that were experimentally known to be exceptionally stable and tightly bound together, just as the quantum theory of atoms explained the existence of chemically particularly inert elements, the "noble gases." It also made it possible to calculate many of the features of the spectra of gamma rays (electromagnetic radiation like light but of very much shorter wavelength) emitted by excited nuclei, even as Bohr's atomic model explained the spectra of the light emitted by atoms. A subsequent "collective model" then regarded the entire swarm of particles that make up a nucleus as one cooperative whole, capable of stretching, vibrating, and rotating. These cooperative motions led to a quantum mechanical understanding of many other aspects of nuclear spectroscopy.

Even though the fact that the neutron in a nucleus fails to decay does not depend on the presence of many other particles and works already in the nucleus of helium, which has only two neutrons and two protons, cooperative phenomena generally occur primarily in systems that consist of many particles. In the context of nuclear physics, "many particles" usually means anywhere from dozens to about 250. Everyday chunks of matter, on the other hand, consist of some 10^{23} atoms or molecules; they are therefore prime candidates for collective effects, and we have already discussed certain aspects of such systems in Chapter 3. There is a property of matter that we did not touch upon in that chapter, however, because our attention was directed toward the arrow of time: its *phase*.

Phase Transitions

We all know that matter comes in three quite distinct phases: solid, liquid, and gaseous.* Although some minor amount of conversion from one of these states to another occurs at all temperatures—liquids evaporate, solids sublimate—there is one quite distinct temperature at which water freezes and another at which it boil (which is why these two temperatures, at atmospheric pressure, are conveniently used for the definition of the Celsius scale). Moreover, if you leave the flame on under a pot of boiling water, the water does not get any hotter; instead, all the heat from the flame is used for the conversion of the liquid into steam. Water with ice in it cools its environment not by getting warmer but by melting more ice and staying at 0°C. In other words, a mixture of water and steam remains at the temperature 100°C, no matter how much heat flows into or out of it, and a mixture of water and ice similarly stays at 0°C.

In the solid phase, the constituent molecules are relatively close together and, in most cases, ordered in a very regular arrangement which we call *crystals*. In the liquid phase, the molecules are somewhat farther apart and disorderly, and in the gaseous phase, they are much farther away from one another and totally without regular order. A given material is, therefore, generally densest in its solid

*The use of the word "phase" for each of these forms of matter is not to be confused with "phase space." It is a historical accident that the same word appears in both meanings.

phase and least dense in its gaseous phase. In each of these forms, the molecules move back and forth with speeds that depend on the temperature, but in a solid they oscillate about fixed equilibrium positions, while in a gas or liquid they move around freely and often collide with one another. The change of one phase to another is called a *phase transition;* as the temperature shifts, it is sudden and discontinuous rather than gradual, and some physical properties of the material—for example, its density—at the transition temperature jump from one value to another.

There are many other phase transitions of matter besides those between the three phases we are familiar with in everyday life, and they may be harder to detect than melting or boiling. A solid in one temperature range may consist of atoms arranged in one kind of crystal, for instance, with all of them at the corners of cubes. At a certain temperature, they suddenly change to another kind of crystal in which the atoms sit at the corners of tentlike tetrahedrons. Or else some distinct physical property of a material may, at one temperature, suddenly disappear. A case in point is the property of *ferromagnetism.*

Ferromagnetism

The ancient Greeks seem to have been the first to discover that iron has unusual magnetic properties, which is why we use the prefix *ferro,* derived from the Greek word for iron. The permanent magnets we use in the kitchen to attach our kids' drawings to the refrigerator are made of iron or alloys of a small number of other substances. If we heat up such a magnet in an oven, at a certain temperature called the *Curie temperature,* after the French physicist Pierre Curie,* it will suddenly lose its magnetization; we have a phase transition from the magnetic state to a nonmagnetic state. How are we to understand this?

The magnetism of a piece of bulk matter is caused by the fact that each electron in it has a tiny "magnetic moment" and acts like a small magnet itself. (The atomic nuclei contribute almost nothing to the

*Pierre Curie's wife was Maria Slodowska, better known by her married name, Marie Curie, the first woman to hold a professorship at the Sorbonne and one of only two persons ever to have received two Nobel prizes in physics, the second jointly with her husband. Their daughter, Irène Joliot-Curie, shared a Nobel Prize in chemistry with her husband, Frédéric Joliot.

overall magnetization because their magnetic moments are very much smaller than those of the electrons.) In a normal material, these small elementary magnets point randomly in all directions and effectively cancel each other out; only when the material is inserted in a magnetic field do they tend to line up along the direction of the field and produce a noticeable effect. In a ferromagnetic material, on the other hand, the elementary magnets line up with one another spontaneously in domains containing very many of them, so that each of these regions acts like a small magnet itself (small and weak compared with the magnet we may hold in our hand, but huge and strong compared with the elementary magnets that are the electrons). If such a piece of material is put in a magnetic field, the magnetic domains tend to line up the directions of their magnetic moments, and they remain that way even after the external magnetic field is turned off. As a result we have a permanent magnet that produces its own field. When the temperature is raised above the Curie point, the spontaneous magnetization of each domain breaks down and the entire effect of ferromagnetism disappears.

The real problem that had to be understood was why do the electronic magnets line up with one another below the Curie temperature, and why do they fail to do so, not gradually but suddenly, as the temperature crosses the Curie point? The means for the solution of this problem came from quantum statistical mechanics. The most powerful device in its tool chest is a function of the temperature called the *partition function,* a sum of terms each of which is a generalization of the Maxwellian distribution of the velocities of gas molecules, discussed in Chapter 3. Every term in the partition function is an exponential function (the natural growth function defined in Chapter 3) of the form $e^{-E/kT}$, in which k is Boltzmann's constant, T is the temperature, and E is the energy of a quantum mechanical state of the system. Since the number of electrons in a magnetic domain is enormous, the calculation of this energy for every possible state, that is, for every possible orientation of the electronic magnetic moments, is extremely complicated, even with the help of powerful computers. The partition function, if it could be computed, contains all the information about the system that we would ever want. In particular, its behavior as a function of the temperature tells us, by means of known equations of quantum statistical mechanics, whether there is a phase transition, and if so, at what temperature.

When physicists know a mathematical equation that answers important questions in principle but do not know a practical method to solve it or to extract information from it, they often resort to the construction of a simplified model. The idea is to replace the very complicated real system by a simpler one that, we have reason to believe, behaves in a qualitatively similar manner with respect to the questions at issue, even though it may not actually approximate the real system, and that promises to be mathematically manageable. Such models are often still quite complicated and may require great mathematical ingenuity to solve, sometimes evolving into minor areas of mathematical research in their own right. The answers they yield may or may not be relevant to the physical questions that gave rise to their invention in the first place. In some cases these answers may lead to approximately correct predictions; in others they may serve only as analogies or metaphors.

The Ising Model

In the case of ferromagnetism, a very fruitful model was invented in 1925 by the German physicist Ernest Ising. This *Ising model* has kept at least two generations of mathematicians and mathematical physicists busy and fascinated. It is simple enough to lead to relevant answers but not so simple that its mathematics is trivial. The idea is to replace the electronic magnets by phantom magnets situated in a regular array or crystal, each pointing either up or down and represented by the number $+1$ or -1, respectively. They interact with each of their nearest neighbors in the array in such a way that their energy is lower when they are both pointing in the same direction; their energy is also lowered by pointing in the direction of an applied magnetic field, which may be directed up or down. (If one state has lower energy than another, there is a force leading from the second to the first.) This apparently childish up-or-down game is, in fact, a simplified but fair representation of the quantum theory for the electronic magnets. What makes even this simplified model very difficult to study is the fact that so long as the number of magnets is finite no discontinuous or jumping behavior of the partition function with changing values of the temperature is possible; such discontinuities can occur only for infintely many of them. This implies that even very sharp or almost sudden jumps with T require a very large number of terms to calculate the energy. We again meet a situation that we have encountered several times before: it is necessary to make an idealization to obtain clear mathematical results.

What we are looking for in the Ising model of ferromagnetism is the possible existence of long-range order or long-range correlations. This means that even though each toy magnet in the model interacts only with its nearest neighbors, they will tend to orient themselves in unison, either all up or all down, even when no external magnetic field is applied. It is as if people in a large crowd, imitating their neighbors as they take off or keep on their jackets, would all spontaneously end up in shirtsleeves rather than in groups with and without jackets, which would be typical short-range order. Long-range order is the microscopic analogue of mass hysteria, and ferromagnetism is one of its physical manifestations.

In one dimension, with our phantom magnets all arranged along a straight line at regular intervals, there can be no phase transition to ferromagnetism in the Ising model. It turns out that in one dimension, the long-range order can too easily be destroyed by a simple break at one point, to the left of which all the magnets are pointing up and to the right down. This prevents the existence of a temperature point below which long-range order is an arrangement that is energetically preferred to short-range order.

In two dimensions, however, with the magnets arranged in a regular array in a plane, like matches lying on a table with their heads up or down, the investment of much ingenuity has led to success. Mathematicians and physicists combined their efforts to solve the Ising model and obtained explicit results. They found that there is indeed a phase transition; above the Curie temperature the toy magnets arrange themselves so that there is short-range order only, while below the Curie temperature they spontaneously choose an arrangement with long-range correlations. Furthermore, many of the relevant thermodynamic functions and their behavior near the Curie temperature have been calculated explicitly, from which we may draw certain qualitative conclusions about the behavior of real ferromagnets that tend to agree with experimental results. In three dimensions, which is of course the case of primary interest, no amount of ingenuity has led to a solution, and there appears to be little hope that a solution will be found. Instead, physicists have had to resort to computers to get approximate answers, which are not as valuable as exact results because, after all, they approximate a model that has limited validity anyway, and it is difficult to determine reliably whether there is a discontinuous jump or not by numerical approximations. Other mathematical models of ferromagnetism have also been explored, but none to the same extent as the Ising model.

COLLECTIVE PHENOMENA

Discovery of Superfluidity and Superconductivity

The first person to liquefy helium, the gas that at room temperature makes your voice sound funny if you breathe it into your lungs and that we use to fill balloons, was the Dutch physicist Heike Kamerlingh Onnes. This was a difficult feat in 1908, because he had to cool helium to an extremely low temperature, and cryogenic techniques were not highly developed; helium boils at 4.2° Kelvin, which is about −269° centigrade. Of all the elements, it is the one with the lowest boiling point, and at atmospheric pressure it never freezes. Just as we use water below its boiling point as a coolant for the engine of our car, so Onnes could use liquid helium to cool down other substances. And indeed when he did that with mercury, which freezes at −38.89° centigrade, and measured its electrical conductivity, he observed, as expected, that the lower the temperature, the better it conducted electrical currents. Slightly below 4.2° Kelvin, however, he found to his great surprise that the electrical resistance of mercury did not just gradually decrease but suddenly disappeared altogether.

Thus, in 1911 Onnes discovered the phenomenon of *superconductivity*. There are many other elements, alloys, and compounds now known to be superconducting, some of them with "transition temperatures" lower than that of mercury and others much higher. The discovery, a few years ago, of "high-temperature superconductors," substances with transition temperatures that could be reached by liquid nitrogen at its boiling point of −196° centigrade (rather than requiring the much more costly liquid helium), caused considerable excitement among both physicists and journalists. No substances are known to be superconducting at, or anywhere near, room temperature, but the search is still going on with great intensity. Such materials would be of great technological interest because their conductivity is, so far as can be determined, absolutely zero; a current in a superconducting ring, once established, will persist forever. Superconducting transmission lines for electrical power would be free of losses that are ordinarily caused by their heat-producing resistance.

An additional strange property of superconductors, discovered in 1933 by the German physicists W. Meissner and R. Ochsenfeld, is that they completely expel any magnetic field from their interior; in a sense they are the exact opposite of ferromagnets. It is this peculiarity, now called the *Meissner-Ochsenfeld effect* or simply the *Meissner effect,* which makes a magnet float freely above a superconducting surface and that is being exploited in such projects as very fast trains employing magnets to glide without friction above superconducting rails.

There is, however, another remarkable feature of liquid helium itself, which Kamerlingh Onnes had missed. In 1938 the Russian physicist Peter Kapitsa, and simultaneously the Canadians John Allen and Austin Misener, found that at 2.2° Kelvin helium has a phase transition to a different type of liquid now called helium II, a *superfluid*. Figure 44 shows a graph of the specific heat* of liquid helium as a function of the temperature in the vicinity of 2.2°, called the *lambda point* because of its resemblance to the mirror image of the Greek letter lambda. The discontinuous behavior at the temperature of the phase transition is clearly visible. Below that transition temperature, the liquid becomes an extraordinarily good conductor of heat, more than 200 times better than copper, the excellent heat conduction of which is taken advantage of in copper-clad cooking utensils and some heating pipes.

Helium II is called a superfluid because it appears to have an extremely low viscosity, less than 1/10,000th that of hydrogen gas; it manages to flow easily through slits 1/100,000th of an inch wide, which water would penetrate only very slowly. The almost nonexistent viscosity of this strange liquid makes it perform such unusual feats as creeping up the inside of an upright glass container, flowing down on the outside, and rapidly dripping from the bottom. And yet, if a paddlewheel is immersed in helium II and turned, which is the usual way of measuring the viscosity of a fluid, it does offer considerable resistance, as though it did not have such a low viscosity after all.

The explanation for the astonishing phenomenon of superfluidity was found within a few years, but for the equally astounding superconductivity the explanation was not found for over forty years.

Superfluidity

Three crucial ideas led to an understanding of the behavior of superfluids. The first came from the Hungarian-American Laszlo Tisza, who suggested that helium II may be thought of as consisting of two interpenetrating fluids moving quite independently of one another.

---

*The specific heat of a substance is the amount of heat required to raise the temperature of one gram by one degree.

COLLECTIVE PHENOMENA

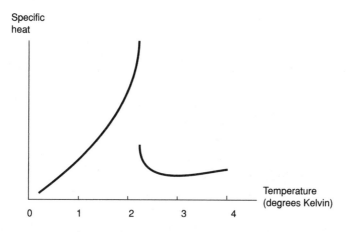

FIGURE 44. The specific heat of helium as a function of the temperature. At the lambda-point it has a discontinuity and its slope is infinite.

One of them is like a normal liquid, and the other has no viscosity at all. This "two-fluid model" would explain the discrepancy between the two measures of its viscosity: the superfluid component is the part that seeps through very fine orifices and creeps up on the walls of containers, and the normal component is responsible for the resistance offered to a paddlewheel.

The other two ideas that formed the basis for an understanding of the superfluid originated from quantum mechanics, and they imply that superfluidity is a macroscopic manifestation of quantum phenomena. In contrast to almost all other instances of quantum effects, we need no microscopes or other fancy instruments to make the radical change from nineteenth-century physics visible; what we see here with our naked eye cannot be explained by classical physics.

The first of the two essential quantum ideas is based on the fact that in the quantum world, particles of the same kind are not just similar to each other but are in a fundamental way *identical*. There is, as a matter of principle, no way of tagging them in order to trace their course. This indistinguishability of particles of the same kind has the consequence that their statistical behavior in the aggregate is not the same as it would be if we could tell one from another. In 1923 the unknown young Indian physicist Satyendranath Bose, after whom bosons were later named, had sent a paper to Einstein that pointed this out for the particular case of light quanta; Einstein both arranged for the publication of Bose's idea and also generalized it. The new statistics came to be called *Bose–Einstein statistics.*

Fundamental particles generally fall into two distinct classes, mentioned already in Chapter 8: *bosons,* which do not satisfy Pauli's exclusion principle and obey *Bose–Einstein statistics,* and *fermions,* which do and satisfy *Fermi-Dirac statistics.* One of the most profound results of the quantum theory of fields is that bosons are the particles whose intrinsic angular momentum, or spin, is zero or an integral multiple of Planck's constant, while fermions have spin which is *half* of an odd integral multiple of Planck's constant. Both the neutron and the proton are fermions, and since the nucleus of ordinary helium is made up of a tightly bound system of two protons and two neutrons, we should regard it as a boson, subject to Bose–Einstein statistics. Einstein had argued already in 1924, when he generalized Bose's idea about the statistics of photons, that gases made up of molecules subject to these new statistics should behave peculiarly at very low temperatures, namely, that all their molecules should have a tendency to be concentrated in the same lowest possible energy state. This phenomenon is now called *Bose–Einstein condensation,* and Einstein predicted that it should lead to an unusually low viscosity for helium gas. It was this effect of Bose–Einstein condensation that the German physicist Fritz London recalled in 1938 to help us understand the newly discovered superfluidity of helium II, though this suggestion was not well received at first.

There is, however, a rare isotope of helium called helium 3 (to distinguish it from the more common isotope called helium 4), whose nucleus is made up of two protons and one neutron; it therefore behaves as a fermion and thus obeys quite different statistics. Since the atoms of helium 3 are not subject to Bose–Einstein condensation, London predicted that its low-temperature behavior should be entirely different from that of helium 4, and indeed helium 3 was found not to become superfluid even at temperatures very much lower than 2.2° Kelvin, the transition temperature of helium 4. Some twenty years later, helium 3 was discovered to become superfluid at a temperature about 1/1,000th of the absolute temperature of the superfluid transition of helium 4, with quite different properties, requiring separate explanation. London's concept that Bose–Einstein condensation plays a role in the superfluidity of helium 4 is now recognized as an important component of the correct explanation.

The third part of the understanding of superfluidity came from the

Russian physicist Lev Landau, a man who played a significant role in the modern development of physics in the Soviet Union.* He won the Nobel Prize in 1962 and in the same year suffered an automobile accident that left him physically and mentally disabled until his death in 1968. In 1941 Landau made the following crucial contribution to our understanding of the superfluid phase of helium II.

If we take the Bose–Einstein condensate at zero temperature as a basis, with all the atoms in the same lowest possible energy state of the fluid as a whole, having therefore long-range order, we know there will be more and more vibrations of the atoms as the temperature rises. An application of the quantum theory to these vibrations, which are the same as the molecular vibrations we perceive as sound (see Chapter 5) and which are also analogous to the electromagnetic field oscillations that constitute light, has the result that the energy appears in the form of discrete quanta. In the case of light the quanta, as we saw, are the photons, and in the case of molecular oscillations they came to be called *phonons*. We may therefore picture the superfluid at some nonzero temperature as consisting of a background fluid of atoms all in the lowest state, and a "gas" of phonons, particlelike excitations whose number is not fixed. Now these phonons do not have the same relation between energy and momentum that ordinary particles have. Whereas for everyday objects the energy of motion is proportional to the square of the momentum, for slow phonons the energy is proportional to the first power of the momentum and thus, for small velocities, rises much more rapidly with increasing momentum. At a given low momentum, a phonon has much more energy per unit of momentum than an ordinary particle (see Figure 45).

Imagine, then, the background fluid—the boson condensate—moving through a thin tube. We would expect the collisions of its atoms with the molecules of the wall to slow it down and thus produce the effect of viscosity. Let us look at this from a point of view in which the fluid is stationary and the tube moves. Viscosity would manifest itself if the tube's motion is slowed down by collisions between its molecules and the atoms of the condensate, thereby losing momentum while creating new phonons; momentum being conserved, the

*Landau may be regarded as one of the half dozen physicists of this century who, apart from their own contributions, were most influential through their students or associates; the others, arguably, were Ernest Rutherford, Niels Bohr, Paul Ehrenfest, and Arnold Sommerfeld in Europe, and Robert Oppenheimer in the United States.

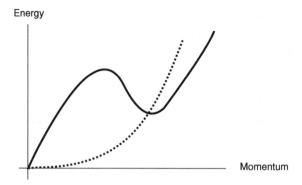

FIGURE 45. The functional relation between the energy of motion and the momentum of a phonon (solid curve) as compared to that of an ordinary particle (dashed curve).

loss is picked up by the newly created phonons. However, energy is also conserved, and phonons have to carry much more energy for each unit of momentum they pick up than the tube molecules lose. Therefore, the slowly moving tube cannot create new phonons and, as a result, will not be slowed down! Seen from the vantage point of the stationary tube, this means the liquid will not be slowed down, which implies that the fluid has no viscosity.

The two-fluid model is now also comprehensible. What seeps through narrow slits is the background liquid—the boson condensate in the ground state—leaving behind all the phonons. This, of course, implies that the residue has higher temperature than it had at the beginning because it has a higher fraction of phonons, an experimental observation that had been puzzling. Similarly, we have an easy explanation for the greater viscosity exhibited against the turning of a paddlewheel. The paddles have to move both the condensate and the phonons, and it is the phonon gas that produces the resistance. Indeed, Kapitsa showed, by heating helium II and thus raising the pressure of the phonon gas by creating more phonons, that this gas can rotate a small vane like a windmill.

We have seen, therefore, that the explanation of all the strange phenomena exhibited by the superfluid helium II rests on two quantum effects: the long-range order implied by the Bose–Einstein condensation of its molecules, and the quantization of its mechanical excitations in the form of phonons with an unusual energy-momentum relation. Let us now turn to superconductivity.

Superconductivity

As a first step, we have to understand how a normal metal wire conducts an electric current and what causes the heat-producing resistance. A metal consists of a regularly arranged crystal of ionized atoms in more or less fixed positions. These atoms have lost one or more of their outer electrons, which are free to move around at will. Of course, there is no reason for them to move in one direction in the wire rather than the other, and they will normally meander around randomly. When an electric field is applied, however, as it is when the wire is connected to a battery, the electrically charged electrons experience a force that induces them to move along the wire, which means that an electric current has been established that would move freely if it were not for the presence of some obstacles. First of all, the crystalline arrangement of the metal ions may contain "impurities" that create hurdles to the moving electrons. Second, unless the wire is at zero temperature, the ions will vibrate, which, as we have seen, translates itself quantum mechanically into the presence of phonons that act as obstacles. Collisions between these phonons and the impurities on the one hand and the moving electrons on the other obstruct the free flow of the current, producing electrical resistance in the wire. The higher the temperature, the more phonons there are in the wire and, consequently, the higher the resistance; conversely, the colder the wire, the smaller the number of phonons and the lower the resistance. That, however, does not account for the complete disappearance of resistance when the material becomes superconducting. The explanation (usually referred to as the *BCS theory*) of this phenomenon was furnished by the collaboration of three American physicists: John Bardeen, Leon Cooper, and John Schrieffer, for which they shared the Nobel Prize in 1972. (It was Bardeen's second; he had received his first for work that led to the invention of the transistor.)

Recall the mechanism leading to superfluidity—the long-range correlations between helium atoms produced by their Bose–Einstein condensation in the ground state. However, electrons are not bosons but fermions, and thus subject to Pauli's exclusion principle; they cannot all condense in the ground state but have to fill the available energy states up to a certain maximal level called the *Fermi surface*. What, then, produces a highly correlated state of the electrons that might explain why they act, in effect, like a superfluid? It is their mutual interaction that is responsible.

The electrons exert two kinds of forces upon each other: they repel one another because they all carry the same positive electric charge, and, in addition, they interact with each other via the phonons. The mechanism here is completely analogous to that which we discussed in Chapter 4, namely, the way in which quantum field theory pictures electromagnetic forces in terms of "virtual" photons: in this case the virtual particles are phonons, emitted by one electron and absorbed by another. We may also think of this interaction as being produced by a change one electron causes in the exact position of the ion lattice, and the consequent force on another electron exerted by that change. The combination of these two kinds of forces acts between any two electrons whose energy is not far below the Fermi surface, but it turns out to be particularly powerful between two electrons, one of which has its spin and momentum oriented in the opposite direction of the other (in the absence of a current-causing electric field), producing a strong correlation between pairs of electrons of that particular kind, called *Cooper pairs*. When the current is flowing, the total momentum of each of these pairs (the sum of the momenta of each of the two), instead of being zero, has a finite value in the direction of the flow.

The upshot of this tight embrace between pairs of electrons is that we now have again, as we had for the atoms of helium II, a highly coordinated state of the conduction electrons near the Fermi surface, and this coordinated state is such that it cannot be changed without the expenditure of a small but finite amount of energy. The conduction electrons whose energy is far below the Fermi level, of course, require at least as much energy as is needed to get them to the Fermi surface because all the lower levels are filled and the Pauli principle prevents them from absorbing any lesser amount. Therefore, if one of the electrons has a minor collision with a phonon or with an impurity in the crystal, nothing can disturb the rigid lockstep of the column of electrons making up the current. In contrast with the result of such a collision in a normal metal, which would change the momentum and energy of the colliding electron and thus impede the current, no hindrance of the flow is possible in the correlated electronic state of a superconductor unless the collision is powerful enough to overcome the gap that exists between the energy of the coherent flow and the state in which at least one of the Cooper pairs has been torn apart. This is what accounts for the fact that, so long as the material is in this special coordinated phase, it cannot offer any resistance to a current. When the temperature is raised, of course, the

COLLECTIVE PHENOMENA

correlations eventually will be broken up all at once by energetic collisions with phonons, and the superconductivity will disappear.

The BCS theory also explains the Meissner effect. The existence of a quantum mechanical state of particle correlations that range over macroscopic lengths rather than stretching over a few interparticle distances only, led London to the derivation of an equation between the current in a superconductor and whatever magnetic field may exist there. This equation, together with one of Maxwell's equations, implies that near the surface inside the conductor the magnetic field must diminish in strength extremely rapidly, with the result that—beyond a tiny "penetration depth"—it is essentially zero: the magnetic field is expelled from the interior of the superconductor.

So the understanding of ferromagnetism, superfluidity in helium 4, and superconductivity is based on a combination of quantum mechanics and long-range correlations. All three may be regarded as true macroscopic manifestations of quantum phenomena that are fundamentally cooperative effects.

The concepts we have discussed in this chapter are truly collective explanations, in which the whole in a real sense exceeds the sum total of its individual parts. They differ in this respect from all the other ideas explored earlier in this book. In the course of our exploration we encountered not only unstructured states of matter, such as gases and liquids, but also highly structured solids in the form of crystals. Such regular arrangements of molecules or atoms play an important role in physics and chemistry. In our final chapter we shall turn to the consideration of *symmetry* as a guiding principle, whose mathematical formulation has formed a powerful tool not only for the description of such neat arrangements of solid matter but for many more abstract and general purposes as well.

10

THE BEAUTY AND POWER
OF SYMMETRY

The intellectual contemplation and visual realization of symmetry has a strong esthetic appeal that has exerted its influence in many areas of thought and art over the centuries. Indeed, symmetric balance has at times been considered synonymous with beauty, and beauty is a powerful motivating force not just in art but in science as well. At the same time, subtle violations of strict symmetry have also been valued, sometimes more highly than the perfection of pure balance. In this chapter, our journey will lead us to an understanding of the uses that have been made of symmetry in our descriptions of the world and to the influence that concept has had. In this context, too, violations of symmetry will turn out to have their own special significance.

To get a feeling for the different kinds of symmetry that are of interest, let us begin by looking at some examples of symmetric designs used by various cultures in the course of history. There are many different species of symmetry to be found, all with their own special esthetic appeal. The simplest variety is the ordinary bilateral reflection symmetry; here are a few instances of this particular kind.*

---

*Some cognitive scientists explain the power of bilateral symmetry over our perception in evolutionary terms: the image of a predator with its eyes fixed on those of

Bilateral Symmetry

Figure 46 shows examples of Renaissance ornaments with reflection symmetry, among others. (What kind of symmetry does the panel on the lower left have?) Figure 47 is Assyrian. Note the configuration of the arms of the two eagle-headed men: the symmetry is not perfect. Figures 48 and 49 are Chinese and pre-Columbian Aztec objects, respectively, with strong bilateral symmetry. Figure 51 shows a musical example, the "Crab Canon" from Johann Sebastian Bach's *The Musical Offering*. The second violin plays the part of the first backwards, so that the whole piece is reflection-symmetric if we regard the second violin as the mirror image of the first.

Such paradigms of bilateral symmetry raise a fascinating problem: is there, in principle, a way of telling left from right other than by convention? If we wanted to communicate to creatures living in another part of the universe what we mean by the right-hand side, how would we go about it? This is what Martin Gardner, in *The Ambidextrous Universe*, calls the Ozma problem. (An attempt, begun in 1960, to listen by radio telescope for messages from civilizations in other regions of our galaxy was called Project Ozma, after the wizard of Oz.) In contrast to up and down or front and back, left and right are conventions not tied to any physical effect (such as gravity, which determines up and down, or locomotion, which determines front and back). Indeed, until some thirty five years ago, it was thought to be impossible to communicate to aliens from another planet what we mean by right and left without physically pointing in those directions. We shall see later why this is no longer so.

Ernst Mach, an influential Austrian physicist of the nineteenth century, relates the great surprise and shock he experienced as a boy when he first learned that a magnetic needle suspended parallel to a wire through which an electric current flows is deflected and turns (see Figure 50). The situation appears to be completely symmetric with respect to the plane of needle and wire, and yet the needle turns out of that plane when the current starts. What makes it turn one way rather than the other? Can this effect be used to tell the people on Alpha Centauri what we mean by left and right? The answer is no, because we have no way of telling them which of the two ends of the needle is the north pole!

its prey has bilateral symmetry, and the instant adrenalin-raising recognition of this
danger signal was as important for survival in the wild as a device for fighter pilots
that detects when an enemy radar has locked onto his plane is today.

THE BEAUTY AND POWER OF SYMMETRY

FIGURE 46. Metal etching designs for typography, from the pattern book by Peter Flötner, 1546.

FIGURE 47. An Assyrian design, ninth century BC.

THE BEAUTY AND POWER OF SYMMETRY

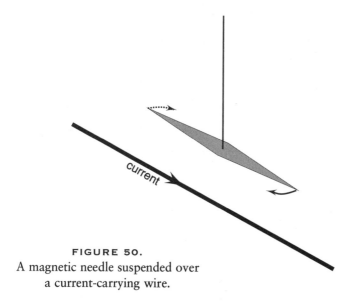

FIGURE 48. Ornamental screen of painted lacquerware from the Warring-States Period of China, 403–221 BC.

FIGURE 49. Aztec ornamental pendant made of wood inlaid with turquoise (now at the British Museum).

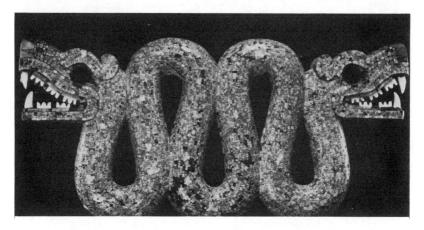

current

FIGURE 50.
A magnetic needle suspended over
a current-carrying wire.

[Crab Canon]

J. S. Bach from "The Musical Offering"

FIGURE 51. The "Crab Canon" from *The Musical Offering* by Johann Sebastian Bach.

THE BEAUTY AND POWER OF SYMMETRY

Translations and Rotations

Of all the many possible symmetries, bilateral reflection symmetry is, of course, the simplest. Instances of translational symmetry by finite steps abound in the art of many cultures. Figure 52 depicts an example from the ancient Greeks, while Figure 53 is from Persian art.

Another frequently encountered symmetry is that of rotations by a given fraction of 360° about an axis. Examples are provided by the Greek amphora shown in Figure 54, the Chinese food container of Figure 55, the Egyptian capitals depicted in Figure 56, and the baptistery of Pisa shown in Figure 57. Nature also contains many designs of this kind. Figure 58 shows examples from among the lower animals; note the octahedron (2), consisting of eight equilateral triangles, the icosahedron (3), consisting of twenty equilateral triangles, and the dodecahedron (5), made up of twelve regular pentagons. Figure 59 shows pictures of snow flakes, with hexagonal patterns. All of these instances actually combine reflectional with rotational symmetry by finite angles. If you leave out the former, you get examples like the swastika or the Pueblo pottery design from Arizona shown in Figure 60. (Notice the broken symmetry of the design near the edge.)

The designs of the great variety of floor tilings found in many cultures usually employ translation, reflection, or rotation symmetries, and in most cases, combinations of them. The tiling shown in Figure 61 was designed by Johannes Kepler, who, in additon to his famous astronomical work, developed a number of tiling theories and many ingenious designs. This pattern has reflectional and rotational symmetry (by 72°) but no translational symmetry. The design depicted in Figure 62, on the other hand, has all three symmetries. The total number of different tiling symmetries is seventeen (which, however, does not mean that there are only seventeen kinds of tiling designs; some recent ones have no symmetry at all). In three dimensions, there are altogether 230 different possible spatial symmetries. However, there are only five regular polyhedra, the so-called Platonic solids: the regular tetrahedron (an equilateral pyramid), the cube, the octahedron, the dodecahedron (whose twelve sides are regular pentagons), and the icosahedron (whose twenty sides are equilateral triangles). If we ask for a combination of rotational and translational symmetry, we obtain a regular lattice, as in a crystal. There are exactly thirty-two such crystal lattices that have some rotational symmetry.

From an esthetic point of view, an imperfect or *broken* symmetry is sometimes preferred. Several of the examples exhibited earlier show such relatively minor asymmetry. The great medieval cathedral at Chartres, shown in Figure 63, one of whose towers differs greatly from the other, is a famous example of a broken symmetry in architecture. The need for humans to show their humility by not aspiring to the perfection that belongs only to God was the purported reasoning behind it. We will return to the role of violated symmetry in physics later on.

The Mathematics of Symmetry

If we want to use the idea of symmetry in science, it is important to capture its essence in mathematical form. First of all, what do we mean by saying that a configuration has a certain kind of symmetry? We imply that if we perform a specific kind of mapping or transformation of all the points in it, the configuration is left unchanged, or "invariant." In the language of physicists, therefore, *symmetry* and *invariance* are used interchangeably.

So we start with the idea of a *transformation* of space, either in two or in three dimensions. Let p be an arbitrary point in space or on a two-dimensional surface; then we define a mapping that takes the point p to some other point p', $p \mapsto p'$. The entire map consists of doing this for every point p, thus defining a transformation $\mathcal{T}$ of the whole space or of the entire surface. For example, the transformation may be a reflection on a plane $\mathcal{P}$, which maps every point p into its mirror image p' with respect to $\mathcal{P}$ (that is, as if $\mathcal{P}$ were a mirror). Or it may be a rotation about an axis by an angle $360°/n$, where n is an integer. We will consider a set of such transformations, chosen to have some specific properties useful for the discussion of symmetry.

Groups

To begin with, we need some simple notation. If the first transformation $\mathcal{T}_1$ is followed by another transformation $\mathcal{T}_2$, then we shall denote the resulting new transformation by $\mathcal{T}_2\mathcal{T}_1$, and we shall call that the *product* of the two. This product, however, has a peculiar property: in contrast with the product of two numbers, for which $9 \times 7 = 7 \times 9$, it need not be true that $\mathcal{T}_2\mathcal{T}_1 = \mathcal{T}_1\mathcal{T}_2$. Performing the transformation $\mathcal{T}_1$ first, followed by $\mathcal{T}_2$, does not necessarily always give the same result as doing them in the opposite order. (We have already seen an example of that in Chapter 4.) For some kinds of

FIGURE 52. An ancient Greek band decoration.

FIGURE 53. Frieze of bowmen from the palace of Darius in Susa, Persia.

FIGURE 54. Amphora found near the Dipilon gate of the Karameikos cemetery in Athens; geometrical period, tenth to eighth century BC.

FIGURE 55. Bronze food container with gold inlay, from the Warring-States Period in China, 403–221 BC.

THE BEAUTY AND POWER OF SYMMETRY

FIGURE 56. Early Egyptian capitals.

THE BEAUTY AND POWER OF SYMMETRY

FIGURE 57. The Baptistery in Pisa, Italy.

THE BEAUTY AND POWER OF SYMMETRY

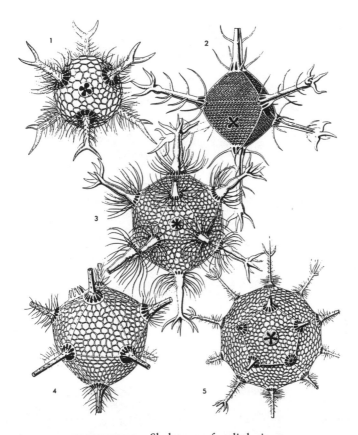

FIGURE 58. Skeletons of radiolarians.

FIGURE 59. Snowflakes.

FIGURE 60. Pueblo pottery design.

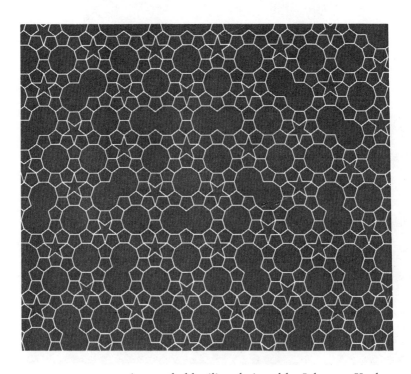

FIGURE 61. A remarkable tiling designed by Johannes Kepler.

FIGURE 62. Binding of the *Kuran of Oeldjaitu* manuscript,
1313 AD, at the Kedivial Library in Cairo.

FIGURE 63. West facade of the cathedral of Chartres,
twelfth to thirteenth century AD.

THE BEAUTY AND POWER OF SYMMETRY

transformations the order does not matter, for others it does. If $\mathcal{T}_2\mathcal{T}_1 = \mathcal{T}_1\mathcal{T}_2$, we say that the two transformations commute. (A set of transformations all of whose members commute is called *abelian*, after the Norwegian mathematician Niels Henrik Abel.)

On the other hand, the so-defined multiplication is *associative*; this means that $\mathcal{T}_3(\mathcal{T}_2\mathcal{T}_1) = (\mathcal{T}_3\mathcal{T}_2)\mathcal{T}_1$. It makes no difference whether we perform the transformation that combines $\mathcal{T}_1$ with $\mathcal{T}_2$ and follow it up by $\mathcal{T}_3$, or whether we perform first $\mathcal{T}_1$ and then follow it up by the transformation that combines $\mathcal{T}_2$ with $\mathcal{T}_3$.

We now want to form the set of all the transformations under which a given configuration is invariant, that is, does not change form. If both $\mathcal{T}_1$ and $\mathcal{T}_2$ are in the set we are forming, the transformation $\mathcal{T}_2\mathcal{T}_1$ should also be in it: if a configuration is invariant with respect to the two transformations $\mathcal{T}_1$ and $\mathcal{T}_2$, it must also be so under the combined transformation $\mathcal{T}_2\mathcal{T}_1$. The transformation that undoes the mapping $p \mapsto p'$ and maps $p' \mapsto p$ is called the *inverse* of the first, written $\mathcal{T}^{-1}$. If $\mathcal{T}$ is in the set, we want $\mathcal{T}^{-1}$ also to be in it. Again, this is clearly needed because symmetry under $\mathcal{T}$ surely implies symmetry under $\mathcal{T}^{-1}$. Finally, we include in the set also the *identity transformation,* which is the transformation that does nothing, $p \mapsto p$, because every configuration is unchanged under this map. A set with these four properties is called a *group*.

We are, therefore, forming a group of transformations, and a configuration has a certain symmetry if it is unchanged, or *invariant*, under all the transformations in that group: this is the *symmetry group* of the configuration.

For example, the transformation $\mathcal{R}$ that describes the bilateral reflection symmetry in the plane which was shown in Figures 46–50 is a mirror-reflection on a vertical line $\mathcal{L}$. This transformation takes every point p into a point p' that lies on the same horizontal line on the opposite side of $\mathcal{L}$ as far from it as p. Therefore, the group has only two members, the identity transformation, which does nothing, and the reflection $\mathcal{R}$ on $\mathcal{L}$. It is clear that two reflections take us back to the original point, so $\mathcal{R}^2 = I$ if we write $\mathcal{R}^2$ for $\mathcal{R}\mathcal{R}$ and I for the identity transformation.

For another example, let $\mathcal{C}$ be the transformation in the plane that takes every point p into a new point p' which is rotated about the coordinate origin by an angle a. If $a = 360°/n$, where n is an integer, then repeating $\mathcal{C}$ n times brings every point back to where it started, $\mathcal{C}^n = I$. Thus we have an abelian group of n members (one says, of *order n*). If no such integer n exists, then the group is of infinite order.

A swastika is an example of a figure that has the symmetry of the rotation group of order four; it is, however, not symmetric under reflections across an axis.

Suppose, next, that we add the reflection across the vertical line through the origin to the rotations. We now have a rotation-reflection group, or *improper* rotation group (in contrast to the *proper* rotation group, which does not include such reflections). This group is not abelian. Take a point on a ray that makes the angle a with the horizontal; a rotation by b puts it on a ray that makes the angle $a + b$ with the horizontal, and a subsequent reflection on the vertical line through the origin puts it on the ray at $180° - a - b$. But if the reflection is done first, it goes to $180° - a$, and the subsequent rotation takes it to $180° + b - a$ (see Figure 64).

Note that the picture shown in Figure 47 does not have reflection symmetry in two dimensions, as mentioned earlier, but if it is regarded as a picture of a three-dimensional scene, its upper part has rotational symmetry by $180°$ about the vertical axis.

In general, the structure of a group is completely determined if we write down the results of all possible multiplications of any two of its members. This is a multiplication table, analogous to the multiplication table we all learned in elementary school. For example, consider the permutations of three objects. By the permutation {231} we mean that the object in the second place should go first, the object in the third place should be second, and that in the first place should be third. The transformation {231} performed after {132} is the same as {321}, or {231}{132} = {321}. All of these products may be assembled in a 6 × 6 multiplication table, which contains the entire structural information about the group of permutations of three objects. There are many other groups of order 6 that have the same multiplication table, which are therefore essentially identical and have the same relevant properties.

Why Does Symmetry Matter in Physics?

The symmetry groups of greatest interest in physics are those consisting of spatial transformations in three dimensions, although symmetry groups in two-dimensions and groups of transformations in abstract mathematical spaces are also pertinent. Particularly in the physics of solid materials and in chemistry, three-dimensional transformation groups play important roles for the description of crystals, whose molecules are arranged in a regular lattice, an arrangement

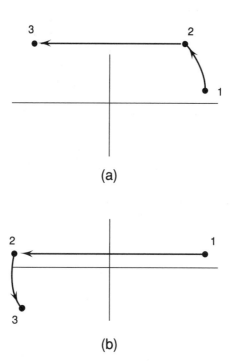

(a)

(b)

FIGURE 64. (a) A rotation followed by a reflection, and
(b) a reflection followed by a rotation.

that usually has both translational and rotational symmetries. That there are exactly thirty-two *point groups* (a fact mentioned earlier and proved by Fredrick Seitz in 1934), that is, groups of transformations in which one point of the crystal is kept fixed but which are compatible with translational symmetry, puts a very important restriction on the possible kinds of crystals that can exist in nature. (It is interesting but humbling to physicists that the fundamental reason why many solids arrange themselves in the form of crystals, though known to be based on quantum mechanics, is not yet fully understood.)

Suppose, now, we know that a physical system has a certain symmetry group. What consequences can we expect? As we have discussed, microscopic physical systems—systems made up of microscopic particles—are described by quantum mechanics. There are many properties of a given system we will want to explain on the basis of the properties of its constituents and the equations of quantum mechanics; one of the most important of these is its energy. The

THE BEAUTY AND POWER OF SYMMETRY

basic tool for such a calculation is the Schrödinger equation, which in many practical cases is enormously complicated and very difficult to solve. However, there are certain qualitative things that may be said on the basis of this equation without having to solve it completely.

The Schrödinger equation determines, among other things, the energy that a given physical system can have when it is in a steady state, that is, when it is unchanging in time. The set of these allowed energies is called its *spectrum,* as mentioned in Chapter 5. In many important instances, the spectrum consists of certain specific *discrete* values only; this was the great revolution that separated quantum mechanics from classical physics. Whereas classically, atoms could have any energy but could not remain in a steady state, according to quantum mechanics they can exist and remain in certain states with discrete energies only, and the stability of these states, together with their properties, explains all of chemistry.

The Schrödinger equation alone, however, does not necessarily determine the state of a given system that has a specified allowed energy *uniquely;* there may be more than one possible state of the system with that energy. If this is so, the energy is called *degenerate.* If there are *n* states with the same energy, it is called *n*-fold degenerate, and this degree of degeneracy is an important property of the system. If we want to know what parameters are required to specify the state of a given system completely, we need to know the possible amount of degeneracy it can have.

Imagine, then, that we are confronted with a complicated physical system, say, a lead atom with 82 electrons and a nucleus in a given physical environment. To solve the Schrödinger equation would be prohibitively difficult even with today's computers. But we may know its *symmetry group,* and it turns out that that alone allows us to make very specific predictions about the possible degeneracies it can be expected to have. The tools for this are the theory of groups and what is called *representation theory.* In fact, each of the possible energy levels of the atom can be classified according to universal rules derived by means of the theory of group representations—for this we need no complicated computation or detailed knowledge of all the forces involved; it is sufficient to know the group of its symmetries. The "normal" degeneracies of the various energy levels are then completely determined and can be calculated relatively easily; there may also be some additional degeneracies in the actual system—these are called "accidental."

Rattled Symmetry

A second question, the significance of which may perhaps be appreciated more directly, arises when the system is subjected to a small perturbation that destroys its symmetry: we say its symmetry is *broken*. Such a perturbation in an atom, for example, may be caused by an electric field. The atom will then not remain in its original steady state forever but will make a transition from one energy level to another by emitting or absorbing a photon of visible light. We will want to know the probability for a specific transition to occur, which determines the intensity of the light that a piece of material made up of such atoms will emit, or the amount it will absorb. If certain kinds of transitions turn out to have probability zero, that particular kind of light will never be seen: such a transition is called *forbidden,* and these strictures are called *selection rules*. They form the theoretical basis of the experimental science of spectroscopy. One of the powers of group theory is that it permits the relatively simple calculation of such selection rules on the basis of symmetry considerations alone, using the group-theoretical classification of its energy levels, without having to know the detailed structure of the system. We will return to this point later with specific examples.

Continuous Symmetries

All the symmetry groups mentioned so far have a finite number of members, that is, their order is finite, so that each member can be labeled by a discrete index. But there are other groups whose members have to be labeled by continuous parameters. They are called *Lie groups* after the Norwegian mathematician Marius Sophus Lie. Important examples of these are the rotations and translations, with no restrictions on the angles of rotation or on the amount of translation. Pictures with such symmetries, of course, are rather boring. The only surface that has complete rotational symmetry is a sphere, and a picture with complete translational symmetry in some direction can have variations in colour or brightness only in the lateral direction. However, these symmetry groups play an especially important role in physics because many idealized physical systems have them.

The fact that the members of Lie groups are labeled by continuous parameters implies that if one or more of these parameters is changed by a tiny amount, we obtain again a member of the group. For the groups of interest, the members change continuously with the parameters. We can then study these infinitesimal changes rather than the group itself, which in many cases is simpler. The collection of such small changes in the group elements when the parameters are changed by small amounts is called the *Lie algebra* that *generates* the Lie group. The properties of one determine those of the other. For example, instead of investigating how a system changes under a finite translation from one position to another, we may more simply study its behavior under a tiny shift in some direction. Or, instead of probing its properties when it is arbitarily rotated, we may test it for a tiny rotation about an arbitrary axis. The results of the latter determine the former.

Perhaps the most basic physical system from the point of view of our own existence is the atomic system consisting of a charged particle (electron) in the electric field of an oppositely charged nucleus—a hydrogen atom. This system has certain degeneracies that are "normal" because of its rotational symmetry group, and it also has some additional degeneracies that are, from the point of view of the rotation group, accidental. However, it turns out that this system also has another less obvious kind of symmetry which is equivalent to a rotation in four spatial dimensions, and when this group is taken into account, all the accidental degeneracies of the system become normal. Since the entire periodic table of the elements—the foundation of all of chemistry—is based on fitting atomic configurations into the available spaces with allowed energies, it may be argued that the theory of symmetry groups is one of the cornerstones of chemistry.

The Role of Invariance in Classical Physics

Let us now make a little detour into classical physics as formulated by Newton, Maxwell, Lagrange, Laplace, and Hamilton (see Chapters 2 and 4).

Recall that the motions of particles and extended bodies and the variation of electromagnetic fields with time as well as from one point to another in space are determined by a combination of differential equations, boundary conditions, and initial conditions. The differential equations embody the general laws of motion for the particles or bodies and the general field equations; the initial conditions and boundary conditions contain all the contingent information. For example, if we want to know the trajectory of a baseball, we need its equation of motion—that is, Newton's equations of motion with information about the gravitational field and, depending on how

THE BEAUTY AND POWER OF SYMMETRY

precisely we want to calculate the trajectory, about the air pressure, wind velocity, and so on. But we also have to specify where it was hit, in which direction, and with what speed—we have to give the initial conditions. For another example, in order to calculate the electromagnetic field in a given region, we need to know not only the Maxwell equations, that is, the differential equations which they obey, but also how strong the electric and magnetic fields are on the surface of the region; those are the boundary conditions. When we say that Newton's equations determine the orbits of the planets around the sun, it is understood that in order to determine an orbit uniquely, we also have to know the initial conditions, that is, where the planet stood at one time and what its velocity was. All the planets obey the same equations of motion, but they satisfy different initial conditions.

Because of the perfection of the circle, which is the only curve in two dimensions with complete rotational symmetry, planetary orbits used to be pictured as circles; as God's creations they had to be perfect. This made Kepler's model, in which they are ellipses, particularly repugnant because they lacked the divine purity of the circle. From our present point of view, we would say the problem with this kind of argument is that, even if the law embodied in a differential equation has a symmetry considered desirable or beautiful, the solutions of the equation need not have the symmetry, because the contingent boundary or initial conditions do not have them. Newton's equation of motion for a planet around the sun has, in fact, perfect rotational symmetry and is thus as beautiful and pure as we could wish. But that does not make all the trajectories circles—most of them are ellipses. This simply means that our admiration of beauty has been shifted to a more abstract level. A seeker of divine perfection or sublime beauty should not look for it in the accident-prone reality of the world but in the underlying laws that govern nature.

Symmetries and Conservation Laws

If we are searching for a law that embodies certain symmetries, how do we actually go about finding it? Or conversely, how do we determine whether a given law has certain invariance properties? A very elegant and powerful formulation of Newton's equations, which much facilitates the answer to such a question, was given by Lagrange. We start with a single given function that contains all the information about the forces, the masses, and so on, and then ask:

what motion of the objects or particles from a given initial configuration and time to a final configuration and time is such that this function, called the *action,* has the smallest possible value? The unique motion that minimizes the action is, in fact, the one that follows Newton's laws: this is called *Hamilton's principle of least action.* The same kind of formulation is also possible with Maxwell's equations and the Lorentz equations that determine the motion of electrically charged particles in an electromagnetic field. Thus, once we know the action function, the content of the equations of motion and of the field equations is determined. This is very convenient for the consideration of symmetries because in order to determine whether the equations of motion of a system are symmetric, we need only test if the action has that property, which usually is much simpler.

Let us assume that a given physical system has a certain symmetry embodied in a Lie group or as generated by a Lie algebra. What are the consequences of this? A very fundamental theorem by the German mathematician Amalie Emmy Noether asserts that to every symmetry or invariance of this kind there corresponds a *conservation law* for the system. Such conservation laws are among the most important consequences of physical laws of motion and field equations; in exceptional cases they even allow us to determine the behavior of a system completely. As we saw in Chapter 2, the conservation laws put stringent restrictions on the possible motions of a system; their power resides in their absolute character and generality. The best known of them is, of course, the law of conservation of energy.

Conservation of Energy

A complicated history (see Chapter 3), featuring the names of Helmholtz, Joule, Mayer, and Kelvin, lies behind the law of conservation of energy and its universal applicability to thermodynamics, physiology, and biology. For our present purpose we are not so much interested in that wider aspect of the law. What is important for us is that for every physical system of particles and fields in interaction with each other there exists a quantity called the energy which is conserved, that is, which does not vary with time. Conservation of energy can be deduced directly from Newton's equations of motion together with the applicable field equations. It is a remarkable fact that this conservation law is a simple consequence of Noether's theorem together with a postulated symmetry property: if we assume that the system is invariant under translations in time, the conservation law that is implied by Noether's theorem is that of energy.

What does *invariance under translations in time* mean? It is the simple notion that if we set an isolated system in motion today at 5 p.m. and examine it tomorrow at 6 a.m., and we compare the result with what happens if we start it *in the same way* at 7 p.m. and examine it tomorrow at 8 a.m., the two final configurations will be found to be the same. There is no built-in clock that makes the system behave differently today as compared to yesterday or a hundred years from now; this is what is called *time-translational invariance*. Noether's theorem tells us that if the system has this invariance property, its energy will be conserved; if it does not—for example, if some internal mechanism in it "runs down"—its energy will not be conserved. This argument can in fact be turned around: in any given theory, the quantity that is conserved by virtue of its being invariant under time translations is, by definition, the energy. That is the most unambiguous way of identifying what we mean by the energy in any newly proposed theory.

Conservation of Momentum and Angular Momentum

There are other invariance aspects of isolated systems that we ought to examine. One of these is invariance under *spatial* translations. By this we mean that it makes no difference whether the system is set up, put in motion, and later checked, all in a laboratory here or in a laboratory a thousand miles away, or even in another galaxy—the results will be the same. Of course, such an invariance does not hold for all physical systems: we must be sure that the system is *completely isolated*. If the graviational field of the earth intrudes on it here, its behavior will be different on the moon. But an isolated system can be expected to be *invariant under spatial translations*. If this were not so, we could never discover a general law because its validity could never by verified in any laboratory other than the one in which it was first found. (The same, of course, holds for time translations: a law that is not invariant under time translation, if discovered today, could not be checked tomorrow.) It is worth noting that these assumptions hold at best for idealized systems; real physical systems are never completely isolated.

According to Noether's theorem, there ought to be a conserved quantity corresponding to this invariance under a shift in space. Since the translation could be in any direction, the conserved quantity should have a direction as well as a magnitude; it should be a *vector*. This conserved quantity, it turns out, is the total *momentum* of the system. In ordinary mechanics, the momentum of an object is simply the product of its mass times its velocity. The law of conservation of momentum is familiar to us in everyday life. In a head-on collision between a small car and a trailer truck, it is the car that gets totaled and not the truck because conservation of the total momentum of both the truck and the car forces the latter to experience a much greater change in velocity (thus, a more destructive deceleration) than the more ponderous truck. Since velocity is a quantity with a direction, so is the mechanical momentum.

In general, then, one defines the momentum of a physical system as that quantity that is conserved by virtue of the system's invariance under translation. This statement implies that an electromagnetic field (and any other field) also has momentum. Just as the bombardment of a window by pebbles exerts pressure because of the conservation of momentum, so there should be pressure exerted by a light ray impinging on a mirror or on any other object; in the case of light, the photons are the pebbles. There are enthusiasts in several countries who are designing robot vessels with enormous reflecting mylar surfaces called *photon sails* to catch the pressure of sunlight for propelling the ships to Mars, just as Columbus used the power of the wind to sail the ocean.*

Invariance with respect to rotations is a third symmetry we expect an isolated system to possess. If an experiment in a laboratory on the earth performed in the morning is repeated in the same place in the afternoon, we expect to get the same result, even though the rotation of the earth has turned the entire laboratory by a large angle. Experiments done here yield the same results as similar experiments done in Australia. Again, Noether's theorem tells us that this invariance must entail the existence of a conserved quantity with a direction attached to it. (The axis of rotation has a direction.) The conserved quantity is the *angular momentum*.

In ordinary mechanics, the angular momentum of an object with respect to a given point P is the product of its mass multiplied by the square of its distance from that point and by the angular velocity about it; its direction is perpendicular to the plane containing P and

---

*On the other hand, two phenomena that were at one time ascribed to the radiation pressure of light turned out to have other explanations: the turning of the little paddlewheel of a radiometer, and the pointing of a comet's tail away from the sun.

THE BEAUTY AND POWER OF SYMMETRY

the velocity. We are observing the conservation of angular momentum when an ice skater accelerates her twirling by pulling in her extended arms—in order to conserve angular momentum, nature compensates for the decrease in (average) distance of her mass from the center by automatically increasing her angular velocity. For complicated systems and fields, the proper definition of angular momentum is not obvious; the best way is to rely on Noether and define it as the quantity that is conserved if the system is invariant under rotation.

So we have three fundamental quantities in a physical system that are conserved if the system has three basic symmetries: time-translational invariance leads to energy conservation; space translational invariance leads to momentum conservation; and rotational invariance leads to conservation of angular momentum. These implications hold in quantum mechanics and quantum field theory as well as in classical mechanics and classical field theory because Noether's theorem holds there, too. In fact, the connection between symmetry on the one hand and conservation laws on the other is more powerful in quantum theory than in classical theory because in quantum mechanics it can be used in the case of discrete symmetries as well.

Reflection Symmetry in Quantum Mechanics

We began this entire discussion with reflection symmetry. In three dimensions, what you see in a mirror is a world subjected to a reflection on a plane. It is not something you can accomplish by a rotation, as you notice immediately when you try to read mirror-image writing. A left hand is transformed into a right hand, a right-handed screw (which is tightened by turning it clockwise) into a left-handed one. Would we expect the fundamental laws of nature to be symmetric under such a transformation?

This question is not the same as asking whether there are natural systems that have a "handedness." There are many such systems: some sugar molecules turn the polarization of light shining on them to the right and others turn it to the left; in fact, all proteins have a built-in left-handedness. Human hearts are on the left, and the structure of the heart defines a screw sense (see Figure 65). These observed facts, however, do not necessarily imply a "handedness" of the fundamental laws. Remember the distinction between the laws of nature and the contingent initial conditions. Two different kinds of molecule, one the mirror image of the other, may be equally compatible

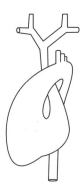

FIGURE 65. Schematic drawing of a human heart;
it has a screw sense.

with the laws of nature; their difference then comes from the accidental initial conditions that are reponsible for forming them. The lack of mirror symmetry built into humans and other animals may have orginated in the accident that formed the earliest life and thence propagated. Indeed, it used to be a firmly believed dogma of physics that *the basic laws of nature are symmetric under reflection.* So the asymmetries observed could have been caused only by accidental initial conditions.

The reflection transformation, together with its repetition, which is the identity (two reflections get you back where you started), forms not a Lie group but an abelian group of order two, as we saw earlier. However, in quantum mechanics, even such discrete symmetry groups lead to "conserved quantum numbers." The quantity conserved by virtue of reflection symmetry is called *parity.* Consequently, every physicist accepted the axiom that "parity is a good (conserved) quantum number" in any natural system, which is equivalent to saying "parity is conserved."

About forty years ago, some particle-collision experiments done by high-energy physicists showed puzzling events for which there were only two possible explanations: either there were two particles, then called *tau* and *theta,* having identical properties, the same mass, and so on, but opposite "intrinsic parities" (that is, they were related to their own mirror images in opposite ways), or else parity was not conserved in nature. Since the second alternative was considered absurd, the first had to be accepted, leading to the "tau–theta puzzle." It made no sense for nature to have produced two particles with exactly equal masses and other properties but opposite parity. (It is

THE BEAUTY AND POWER OF SYMMETRY

an axiom accepted by most physicists that nature does not do silly, stupid, or even superfluous things.)

The gordian knot was cut with the revolutionary proposal made in 1956 by two Chinese-American physicists, C. N. (Frank) Yang of the Institute for Advanced Study in Princeton and T. D. Lee of Columbia University, that parity is in fact not conserved: nature is not symmetric under mirror reflection. Had their argument stopped there, it would, of course, have been neither persuasive nor valuable. However, they bolstered their case by demonstrating that the conservation of parity had not been clearly experimentally demonstrated in the "weak interactions" such as beta radioactivity, which were assumed to be involved in the decays of the tau and theta particles; it had, instead, always simply been taken for granted. And they proposed specific crucial experiments to test whether parity was conserved in radioactive decays of atomic nuclei. These were quickly done by the Chinese-American physicist C. S. Wu and her collaborators at Columbia University, and her results brilliantly confirmed the Lee–Yang theory, thus solving the tau–theta puzzle.

The first reaction of most physicists to the Lee–Yang proposal was incredulity. Wolfgang Pauli, the great discoverer of the neutrino, electron spin, and the Pauli exclusion principle, referring to the nonconservation of parity in the weak interactions, had said before the Wu experiment that he could not believe that "God was a weak left-hander." Lee and Yang won the Nobel Prize in 1957, only one year after their paper, and Pauli ate crow. In a letter to Emil Konopinski in December 1956 Pauli had stated as a rule: "What a modern theoretician says under the title 'universal' consider to be simply nonsense." In February of 1957, after the Wu experiment, he wrote that he accepted Konopinski's proposed addition to the rule that "feelings of symmetry" should be included and also be regarded as nonsense.

Now that we know nature, at the most fundamental level, is not mirror symmetric, we may speculate that observed molecular and biological asymmetries were caused by this basic lack of symmetry. We also have, at last, a solution to the Ozma problem posed earlier in this chapter: it is, in fact, possible to communicate to Martians what we mean by left and right. If they perform the Wu experiment, placing radioactive cobalt in an electric-current-carrying coil and measure the distributions of the emerging beta particles (electrons), they will find a preponderance of emitted electrons in one direction which, together with the direction of the current, defines a left-handed screw (see Figure 66).

THE BEAUTY AND POWER OF SYMMETRY

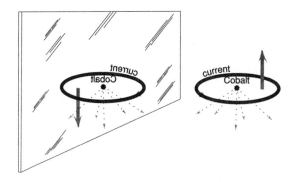

FIGURE 66. Schematic representation of the radioactive decay of a cobalt nucleus in an electric current-carrying coil and its mirror image. The vertical arrows indicate the direction in which a right-handed screw would move if turned in the direction of the current; the dashed arrows show the directions of a preponderance of the emitted electrons.

We may reasonably raise the question whether this breaking of reflection symmetry at the apparently basic level of nature is really *fundamental*. According to more recent theories, the weak interaction, which violates parity, is united with other interactions at an even more basic level that is accessible only at very much higher energies, and at this most fundamental level parity is conserved. According to these theories, the violation of mirror symmetry evolved as a historical accident in the early history of the universe. These ideas, however, have not yet been convincingly confirmed by experiments.

Other Surprises: Violation of Time Reversal Invariance

There were other surprises ahead. For a few years after the parity break, physicists were able to reconcile themselves to the new state of affairs without too much intellectual pain by combining all mirror reflections with another transformation called *charge conjugation,* which replaces all negative electric charges with positive charges and vice versa. Thus it was postulated that the mirror-image of an electron should be taken to be a positron, thereby combining conjugation with reflection; if one did that, the symmetry appeared to be restored by a line of reasoning beginning with a very fundamental conservation law, called the *CPT theorem,* that had been proved to hold in any reasonable theory. It says that the combination of the three transformations—charge conjugation (symbolized by C), spatial reflection

(P, for parity), and time reversal (T)—must be a symmetry of nature. One of the consequences of this law is that the mass of any particle must equal that of its antiparticle; the masses of the electron and positron must be exactly equal, and so must be the masses of the proton and antiproton.

Time reversal? Here is another fundamental symmetry that used to be considered sacrosanct. If a process $A \rightarrow B$ is compatible with the laws of nature, then so must $B \rightarrow A$ be. Or, if you take a movie of some evolving physical system, running the film backwards should also show a possible evolution; in other words, looking at the film alone should not make it possible to determine whether it is running forward or backward. Of course, this must not be a system with friction or other irreversible thermodynamic processes. The second law of thermodynamics does define an "arrow of time" by the increase of entropy, but that is on the basis of statistical properties of many-particle systems (see Chapter 3). At the fundamental level, there was thought to be no "arrow of time."

If this belief is true, that is, if nature has "time-reversal symmetry," then the CPT theorem implies that the combination CP, that is, mirror reflection together with charge conjugation, must also be a symmetry. (Our solution of the Ozma problem would then fail if the recipients of our message lived on an antiplanet!) But just as we thought that it was safe to rely on such CP symmetry, things went awry again.

Another experiment on fundamental particles, done by Val Fitch and James W. Cronin in 1964, demonstrated that in fact the combination CP was not always a good symmetry, and thus neither is time reversal; they are broken by a very small amount. The CPT theorem has never been found to be violated and is considered to be on safe ground. Therefore, we have to conclude that nature, at its most fundamental level, has an arrow of time. Running a film of the development of the universe backwards does not show a possible scenario. However, whether the arrow of time provided by this very weak violation of time-reversal invariance is responsible for the arrow of time that is defined by the temporal order of cause and effect (which we saw in Chapter 3 to be connected to the arrow of time defined by the second law of thermodynamics) is a puzzling open question.

The breaking of CP invariance helps to solve another baffling question: why is the world composed entirely of matter, as it appears to be, rather than islands of matter and islands of antimatter in equal amounts? If this is so—and we cannot really be entirely sure because the light and its spectrum emitted by an antistar would be the same as that from a star—how did such an asymmetry ever arise in the history of the universe? According to present cosmological ideas, when the big bang started the universe it was composed entirely of radiation. It then cooled down, creating pairs of particles and antiparticles in equal numbers. A slight breaking of either C or CP symmetry would lead to an eventual preponderance of one over the other, explaining the present state of affairs.*

Symmetry and Its Breaking in Elementary-Particle Physics

Group theoretical symmetry considerations have played a central role in the development of fundamental theories of elementary particles over the last thirty years. In order to understand why this is so, we need only add a small ingredient of the theory of relativity to what we have already discussed. As we have noted in earlier chapters, according to Einstein's special theory of relativity, the energy of a particle at rest is given by $E = mc^2$, where c is the velocity of light in vacuum and m is the mass of the particle. Therefore, if we have a theory of any kind that provides us with the allowed spectrum of energies, this theory also provides us with the numerical values of the masses of the elementary particles. As we saw before, if such a theory has a known symmetry group, then we are in a position to predict the "normal" degeneracies of the energy levels that can be expected. Consequently, we can also determine how many species of particles should have the same mass; such sets of particles are called *multiplets*.

A familiar example is provided by the neutron and proton forming a *doublet*. All but certain specified "quantum numbers" of the members of a multiplet are identical. (In the case of the neutron and proton, the exceptional quantum number is the electric charge: the proton has a positive charge, but the neutron is electrically neutral.) However, the masses of the neutron and proton are not precisely the same; the neutron is slightly heavier. We account for this mass difference in the following way.

Consider an acceptable theory of neutrons and protons, but with all couplings to electromagnetic fields switched off. Such a theory has

---

*That also requires a breaking of the conservation law of *baryons*, which are the heavy fermions such as protons, neutrons, and so on and in which each baryon counts as 1 and each antibaryon as −1. Some recent theories lead to such a breaking, and experiments are under way to check if the proton is, in fact, stable, as mentioned in Chapter 8.

a certain symmetry group called SU(2), and it predicts a two-fold degenerate state that can be interpreted as the doublet consisting of a neutron and a proton—so these two particles have precisely the same mass. Now we switch on the interaction with the electromagnetic field, which is rather weak because the electric charge of the proton is small and that of the neutron is zero. This altered theory will not have SU(2) symmetry, that is, the SU(2) symmetry will be *broken,* but it will be close to one that has it. (This is analogous to some of the pieces of art shown earlier, which were also not *exactly* symmetric, but nearly so.) Therefore, we expect that in this theory the difference between the masses of the neutron and proton should be small, as in fact it is. One says that the neutron–proton mass difference is "of electromagnetic origin" because it would be zero but for the electric charge of the proton.

The only problem with this very plausible scenario is that almost every attempt to implement it quantitatively leads to a proton that is heavier than the neutron! So it is not a trivial matter to make it work. Nevertheless, this idea lies at the basis of most of our present theories to account for the many new particles that have been discovered during the last forty years.

The particles discovered have masses that fall into certain groupings which, with some imagination, can be called "close together." We then envisage a theory with a symmetry group (which may have nothing to do with physical space or time but work in some abstract mathematical space) that predicts degeneracies and accounts for the multiplicities obtained when the masses that are "close together" are set equal. This is the way in which Murray Gell-Mann arrived at his famous *eight-fold way* on the basis of a symmetry group called SU(3) in 1961.

The next step, then, is to account for the fact that the actual masses of these particles are not equal by postulating a theory in which the original symmetry group is broken, but relatively weakly, so that the mass differences are not too large. Of course, the detailed group-theoretical arguments provide us with some other properties that multiplets must have in common, which are also subject to experimental check.

Gauge Invariance

What I have described does not exhaust the reliance of basic physics on symmetry arguments. By postulating an invariance or symmetry property, C. N. Yang deduced the very existence of the electromagnetic field and the way it acts upon, and is produced by, charged particles. The required transformation is called a *local gauge transformation,* to which all quantum fields are subjected. If it is assumed that the field equations are invariant under this gauge transformation, it follows that there ought to exist a field that acts on, and is produced by, charged particles in just the same way as the electromagnetic field. If we further assume that the theory is invariant under Lorentz transformations (which simply means that it satisfies Einstein's special theory of relativity), this electromagnetic field necessarily obeys Maxwell's equations. So the field equations themselves can be generated by symmetry considerations! All of this does not, of course, imply the existence of such a field in reality, but that could be inferred from the assumption that nature avails itself of all beautiful options.

More generally, recall the discussion near the end of Chapter 4, in which I already alluded to the power of an assumption of local gauge invariance in generating field equations, as first shown by Yang and Mills. Thus, all modern field theories rely ultimately on postulated symmetries of nature, and all the fundamental particles and their properties as discussed in Chapter 8 are theoretically generated by such assumed symmetries. If these basic invariances imply a universe that would be tediously uniform and that differs in many important respects from the real world, the fundamental symmetries are taken to be *broken* by relatively small amounts; just as the subtlest esthetics put great value on small violations of perfect symmetry, so much of the beauty of the cosmos we live in resides in the effects of symmetry-breaking. In some proposed theories, in fact, the amount of symmetry violation by the real world is not always very small and may depend on the energy at which it is probed. Ultimate, stark, and pristinely symmetric nature is assumed to reveal itself only at such enormous energies that there is little chance of ever beholding it experimentally. The holy grail may be forever out of reach to mortals.

Here we have our final example of the freedom enjoyed by scientists in their construction of theories: a given set of coherent observational facts without any symmetries may be explicable in more than one way by postulating an underlying "purer" theory with many beautiful symmetries and by then assuming that these are broken in the real world. (Perhaps this makes nature uglier than the ideal but much more interesting, or perhaps it makes it even more beautiful because somewhat wacky; *chacun à son goût.*) By analogy, the shape

232

FIGURE 67. The figure in the center may be regarded as a
distortion of those on either side of it,
which are more regular.

shown in the center of Figure 67 may be regarded as a distortion of
the more symmetrical ones on either side of it. The further removed
from experiments, the more abstract a postulated symmetrical underlying foundation of a theory is, the more freedom the scientist has to
indulge his imagination. This does not mean that the constructed
theory is necessarily wrong or fails to correspond to nature, but it
does imply that, no matter how convincing the beauty of the constructs, others may find equally valid alternatives. We can therefore
readily imagine that a visitor from another planet may come to us
with an entirely different image of the world.

EPILOGUE

In this book we have explored some parts of the elaborate structure of physics built by the imagination of many scientists in the course of the last 400 years. Although this structure has an impressive, if imperfect coherence, it should not be considered as a revelation of the ultimate "truth" about nature.

Science is not holy scripture, nor do its practitioners consider themselves priests protecting a glittering grail, forever unchanging and pure. What drives scientists on is the thirst to *understand* more than to *use* nature, to build rather than to exploit a comprehensible universe.

The future will, no doubt, bring many surprises, revealing some of our present ideas to be flawed or incomplete, but science must be a continuing activity; once its creativity is exhausted, our civilization will crumble and we will return to the dark ages. It cannot be sustained by technological ingenuity alone or by a routine search for and classification of more and more observations and phenomena.

I have endeavored to demonstrate that science, at the most fundamental level, is very far from being merely an efficient enumeration of experimental facts and empirical rules, nor is its structure simply determined by induction from observations. To think of it only as an orderly collection of intriguing and useful bits of information is to misunderstand its cultural value and its fascination altogether. Science is, in fact, an intricate edifice erected from complex, imaginative designs in which esthetics is a more powerful incentive than utility. Beauty, finally, comprises its greatest intellectual appeal.

FURTHER READING

1. Science, Mathematics, and Imagination

COSRIMS, eds., *The Mathematical Sciences: A Collection of Essays.* Cambidge, Mass.: MIT Press, 1969.

Gerald Holton, *The Scientific Imagination.* Cambridge: Cambridge University Press, 1978.

E. P. Wigner, *Symmetries and Reflections.* Bloomington: Indiana University Press, 1967.

2. Chaos and the Ghost of Laplace

I. B. Cohen, "Newton's discovery of gravity." *Scientific American,* March 1981, p. 166.

Stillman Drake, "Newton's apple and Galileo's dialogue." *Scientific American,* August 1980, p. 150.

David Ruelle, *Chance and Chaos.* Princeton: Princeton Unicersity Press, 1992.

R. S. Westfall, *Never at Rest: A Biography of Isaac Newton.* Cambridge: Cambridge University Press, 1980.

3. Time's Arrow

J. M. Blatt, "Time reversal." *Scientific American,* August 1956, p. 107.

R. G. Brewer and E. L. Hahn, "Atomic memory." *Scientific American,* December 1984, p. 50.

Percy Bridgman, *The Nature of Thermodynamics.* Cambridge, Mass.: Harvard University Press, 1941.

B. H. Lavenda, "Brownian motion." *Scientific American,* February 1985, p. 70.

David Layzer, "The Arrow of Time," *Scientific American,* December 1975, p. 56.

4. Forces Acting through Space

B. S. DeWitt, "Quantum gravity." *Scientific American,* December 1983, p. 112.

D. Z. Freedman and P. van Nieuwenhuizen, "The hidden dimensions of spacetime." *Scientific American,* March 1985, p. 74.

George Gamow, "Gravity." *Scientific American,* March 1961, p. 94.

M. B. Green, "Superstrings." *Scientific American,* September 1986, p. 48.

D. M. Greenberger and A. W. Overhauser, "The role of gravity in quantum theory." *Scientific American,* May 1980, p. 66.

John Hendry, *James Clerk Maxwell and the Electromagnetic Field.* Bristol and Boston: Adam Hilger, 1986.

Gerard 't Hooft, "Gauge theories of the forces between elementary particles." *Scientific American,* June 1980, p. 104.

A. D. Jeffries, P. R. Saulson, R. E. Spero, and M. E. Zucker, "Gravitational wave observatories." *Scientific American,* June 1987, p. 50.

William McCrea, "Arthur Stanley Eddington." *Scientific American,* June 1991, p. 92.

J. M. Weisberg, J. H. Taylor, and L. A. Fowler, "Gravitational waves from an orbiting pulsar." *Scientific American,* October 1981, p. 74.

E. T. Whittaker, *A History of the Theories of Aether and Electricity: The Classical Theories.* New York: Humanities Press, 1973.

L. P. Williams, "André-Marie Ampère." *Scientific American,* January 1989, p. 90.

L. P. Williams, *Michael Faraday: A Biography.* New York: Da Capo Press, originally published New York: Basic Books, 1965.

5. Waves: Standing, Traveling, and Solitary

A. D. Dalmedico, "Sophie Germain." *Scientific American,* December 1991, p. 116.

FURTHER READING

N. H. Fletcher and S. Thwaites, "The physics of organ pipes." *Scientific American,* January 1983, p. 94.

Russell Herman, "Solitary waves." *American Scientist,* July-August 1992, p. 350.

Claudio Rebbi, "Solitons." *Scientific American,* February 1979, p. 92.

T. D. Rossing, "The physics of kettledrums." *Scientific American,* November 1982, p. 172.

6. Tachyons, the Aging of Twins, and Causality

R. G. Newton, "Particles that travel faster than light?" *Science,* 20 March 1970, p. 1569.

Julian Schwinger, *Einstein's Legacy: The Unity of Space and Time.* New York: Scientific American Library, 1986.

7. Spooky Action at a Distance

J. S. Bell, "Bertlemann' socks and the nature of reality." *Journal de physique,* Colloque C2, supplement au no. 3, tome 42, March 1981, p. C2-41.

C. H. Bennett, "Quantum cryptography: Uncertainty in the service of privacy." *Science,* 7 August 1992, p. 752.

Jeremy Bernstein, *Quantum Profiles.* Princeton: Princeton University Press, 1991.

D. C. Cassidy, "Heisenberg, uncertainty, and the quantum revolution." *Scientific American,* May 1992, p. 106.

Bernard d'Espagnat, "The quantum theory and reality." *Scientific American,* November 1979, p. 158.

M. C. Gutzwiller, "Quantum chaos." *Scientific American,* January 1992, p. 78.

John Horgan, "Quantum philosophy." *Scientific American,* July 1992, p. 94.

Max Jammer, *The Philosophy of Quantum Mechanics.* New York: John Wiley & Sons, 1974.

J. M. Jauch, *Are Quanta Real? A Galilean Dialogue.* Bloomington: Indiana University Press, 1973.

N. D. Mermin, "Is the moon there when nobody looks? Reality and the quantum theory." *Physics Today,* April 1985, p. 38.

N. D. Mermin, "Quantum mysteries revisited." *American Journal of Physics,* 58 (1990): 731.

Walter Moore, *Schrödinger: Life and Thought.* Cambridge: Cambridge University Press, 1989.

Abner Shimony, "The reality of the quantum world." *Scientific American,* January 1988, p. 46.

8. What Is an Elementary Particle?

John Bahcall, "The solar neutrino problem." *Scientific American,* May 1990, p. 54.

H. Breuker, H. Drevermann, C. Grab, A. A. Rademakers, and H. Stone, "Tracking and imaging elementary particles." *Scientific American,* August 1991, p. 58.

R. A. Carrigan, Jr., and W. P. Trower, "Superheavy magnetic monopoles." *Scientific American,* April 1982, p. 106.

D. B. Cline, "Beyond truth and beauty: a fourth family of particles." *Scientific American,* August 1988, p. 60.

P. Ekstrom and D. Wineland, "The isolated electron." *Scientific American,* August 1980, p. 104.

Howard Georgi, "A unified theory of elementary particles and forces." *Scientific American,* April 1981, p. 48.

Haim Harari, "The structure of quarks and leptons." *Scientific American,* April 1983, p. 56.

Maurice Jacob and Peter Landshoff, "The inner structure of the proton." *Scientific American,* March 1980, p. 66.

K. A. Johnson, "The bag model of quark confinement." *Scientific American,* July 1979, p. 112.

D. J. Kevles, "Robert A. Millikan." *Scientific American,* January 1979, p. 142.

A. D. Krisch, "The spin of the proton." *Scientific American,* May 1979, p. 68.

J. G. Learned and D. Eichler, "A deep-sea neutrino telescope." *Scientific American,* February 1981, p. 138.

L. M. Lederman, "The upsilon particle." *Scientific American,* October 1978, p. 72.

J. M. LoSecco, F. Reines, and D. Sinclair, "The search for proton decay." *Scientific American,* June 1985, p. 54.

N. B. Mistry, R. A. Poling, and E. H. Thorndike, "Particles with naked beauty." *Scientific American,* July 1983, p. 106.

Michael Riordan, "The discovery of quarks." *Science,* 29 May 1992, p. 1287.

M. J. G. Veltman, "The Higgs boson." *Scientific American,* November 1986, p. 76.

Steven Weinberg, "The decay of the proton." *Scientific American,* June 1981, p. 64.

Steven Weinberg, *The Discovery of Subatomic Particles.* New York: Scientific American Library, 1983.

Frank Wilczek, "Anyons." *Scientific American,* May 1991, p. 58.

FURTHER READING

9. Collective Phenomena

George F. Bertsch, "Vibrations of the atomic nucleus." *Scientific American,* May 1983, p. 62.

Robert J. Cava, "Superconductors beyond 1-2-3." *Scientific American,* August 1990, p. 42.

J. G. Daunt, "Liquid helium-3." *Science,* 26 February 1960, p. 579.

R. J. Donnelly, "Superfluid turbulence." *Scientific American,* November 1988, p. 100.

E. M. Lifshitz, "Superfluidity." *Scientific American,* June 1958, p. 30.

O. V. Lounasmaa and G. Pickett, "The ^{3}He superfluids." *Scientific American,* June 1990, p. 104.

B. T. Mathias, "Superconductivity." *Scientific American,* November 1957, p. 92.

N. D. Mermin and D. M. Lee, "Superfluid helium 3." *Scientific American,* December 1976, p. 56.

R. D. Parks, "Quantum effects in superconductors." *Scientific American,* October 1965, p. 57.

Fred Reif, "Superfluidity and quasi-particles." *Scientific American,* November 1960, p. 139.

Fred Reif, "Quantized vortex rings in superfluid helium." *Scientific American,* December 1964, p. 116.

10. The Beauty and Power of Symmetry

Marie-Anne Bouchiat and Lionel Pottier, "Atomic preference between left and right." *Scientific American,* June 1984, p. 100.

T. Goldman, R. J. Hughes, and M. M. Nieto, "Gravity and antimatter." *Scientific American,* March 1988, p. 48.

H. E. Haber and G. L. Kane, "Is nature supersymmetric?" *Scientific American,* June 1986, p. 52.

R. A. Hegstrom and D. K. Kondepudi, "Handedness of the universe." *Scientific American,* June 1990, p. 108.

O. E. Overseth, "Experiments in time reversal." *Scientific American,* October 1969, p. 88.

Tony Rothman, "The short life of Évariste Galois." *Scientific American,* April 1982, p. 136.

Hermann Weyl, *Symmetry.* Princeton: Princeton University Press, 1952.

E. P. Wigner, "Violations of symmetry in physics." *Scientific American,* December 1965, p. 28.

Frank Wilczek, "The cosmic asymmetry between matter and anti-matter." *Scientific American,* December 1980, p. 82.

S. M. Girvin, "Anyons superconduct, but do superconductors have anyons?" *Science,* 4 September 1992, p. 1354.

General Reading

Pierre Duhem, *The Aim and Structure of Physical Theory*. Princeton: Princeton University Press, 1991 (first published 1914).

Helen Dukas and B. Hoffmann, eds., *Einstein: The Human Side*. Princeton: Princeton University Press, 1979.

Freeman Dyson, *Infinite in All Directions*. New York: Harper and Row, 1988.

Albert Einstein, *Out of My Later Years*. New York: Philosophical Library, 1950.

Richard Feynman, *The Character of Physical Law*. Cambridge, Mass.: MIT Press, 1965.

Vernard Foley and Werner Soedel, "Leonardo's contributions to theoretical mechanics." *Scientific American,* September 1986, p. 108.

George Gale, "The anthropic principle." *Scientific American,* December 1981, p. 154.

Owen Gingerich, "The Galileo affair." *Scientific American,* August 1982, p. 132.

Sheldon Glashow, *Interactions: A Journey through the Mind of a Particle Physicist*. New York: Warner Books, 1988.

Gerald Holton, *Thematic Origins of Scientific Thought: Kepler to Einstein*. Cambridge, Mass.: Harvard University Press, 1988.

L. M. Lederman, "The value of fundamental science." *Scientific American,* November 1984, p. 40.

L. S. Lerner and E. A. Gosselin, "Galileo and the specter of Bruno." *Scientific American,* November 1986, p. 126.

Russell McCormmach, ed., *Historical Studies in the Physical Sciences*. Philadelphia: University of Pennsylvania Press, 1971.

Abraham Pais, *Inward Bound: Of Matter and the Forces in the Physical World*. Oxford: Clarendon Press, 1986.

Abraham Pais, *Niels Bohr's Times, in Physics, Philosophy, and Polity*. New York: Oxford University Press, 1991.

Abraham Pais, *Subtle Is the Lord . . .: The Science and the Life of Albert Einstein*. Oxford: Clarendon Press, 1982.

David Park, *The How and the Why: An Essay on the Origins and Development of Physical Theory*. Princeton: Princeton University Press, 1988.

Anthony Zee, *Fearful Symmetry*. New York: Macmillan, 1986.

References

Introduction

Page

2 "In thinking about the history of science": Linus Pauling, "The value of rough quantum mechanical calculations." *Foundations of Physics*, 22 (1992): 830. (My emphasis of "approximate.")

8 "pleasing to the mind": Quoted in Owen Gingerich, "Astronomy in the Age of Columbus." *Scientific American*, November 1992, p. 105.

1. Science, Mathematics, and Imagination

Page

9 "'Fact, fact, fact!'": Charles Dickens, *Hard Times*, p. 8.

12 "In the eighteenth century": Quoted in Freeman Dyson, *The Mathematical Sciences*, COSRIMS, eds., Cambidge, Mass., MIT Press, 1969, p. 105.

13 "I want to know how God": Quoted in A. Zee, *Fearful Symmetry*, New York: Macmillan, 1986, p. xi.

13 "What really interests me": Quoted in Gerald Holton, *The Scientific Imagination*. Cambridge: Cambridge University Press, 1978, p. 281.

14   "What I remember most clearly": Quoted in Zee, *Fearful Symmetry,* from *Einstein: The Man and His Achievement,* ed. G. J. Whitrow. New York: Dover, 1973.

15   "intuition supported by rapport with experience": Quoted in Holton, *Scientific Imagination,* p. 95, from *Ideas and Opinions by Albert Einstein,* trans. and rev. Sonja Bargmann. New York: Crown Publishers, 1954, p. 226. However, "rapport with experience" is my own translation of the German phrase *"Einfühlung in die Erfahrung."*

15   "free inventions of the human intellect": Quoted in Holton, ibid., p. 96, from the Herbert Spencer Lecture by A. Einstein, "On the method of theoretical physics," delivered on June 10, 1933, at Oxford. The continuation is quoted on p. 144.

15   "there is no logical bridge": Also from the Spencer lecture, quoted in Holton, ibid., p. 95.

15   "The process by which we come to form": Quoted in Holton, ibid., p. 109, from P. B. Medawar, *Induction and Intuition.* Philadelphia: American Philosophical Society, 1969, p. 46.

15   "Let the imagination go": Quoted in L. Pearce Williams, *Michael Faraday: A Biography.* New York: Da Capo Press, p. 467.

16   "the physicists have . . . some justification": Quoted in Holton, *Scientific Imagination,* p. 49.

17   "A physicist builds theories": Dyson, *Mathematical Sciences,* p. 106.

18   "The unreasonable effectiveness of mathematics in the natural sciences": Richard Courant Lecture in Mathematical Sciences, delivered by Eugene Wigner at New York University, May 11, 1959; reprinted in Wigner, *Symmetries and Reflections: Scientific Essays.* Bloomington: Indiana University Press, 1967.

21   "The mathematics enables": Dyson, *Mathematical Sciences,* p. 106.

2. Chaos and the Ghost of Laplace

Page
23   "Given for one instant": From the English translation of "Essai sur les probabilités." New York: Dover, 1951, p. 4. (Original published in 1819.)

3. Time's Arrow

Page
65   "smell more": P. W. Bridgman, *The Nature of Thermodynamics,* Cambridge, Mass.: Harvard University Press, 1941, p. 3.

4. Forces Acting through Space

Page

80   "that one body may act upon another": Quoted in E. Whittaker, *A History of the Theories of Aether and Electricity: The Classical Theories.* New York: Humanities Press, 1973, p. 28.

80   "Yet what we have said about these forces": Quoted in Richard S. Westfall, *Never at Rest.* Cambridge: Cambridge University Press, 1980, p. 390.

80   "I don't care": Westfall, ibid., p. 464.

80   "A Frenchman who arrives in London": Whittaker, *Theories of Aether,* p. 29.

80   "It is the language used": Whittaker, ibid.

81   "His doubt was whether he should": Westfall, *Never at Rest,* p. 647.

82   "As argument against the received theory of induction": Quoted in Williams, *Michael Faraday,* p. 298.

83   "If [the lines of force] exist, it is not": Ibid., p. 450.

84   "With regard to the great point": Ibid., p. 454.

84   "When I contemplate gravitation": Ibid., p. 508.

5. Waves: Standing, Traveling, and Solitary

Page

115   "Can one hear the shape of a drum?" M. Kac, *American Mathematics Monthly,* 73, no. 4, part II, pp. 1–23.

121   "I was observing the motion": From J. S. Russell, "Report on waves," British Association for the Advancement of Science Reports, 1844.

121   "If such a heap be by any means forced": Ibid., p. 323.

122   "Having ascertained that no one": Ibid., P. 333.

122   "We accordingly find that a theory": Ibid., p. 334.

7. Spooky Action at a Distance

Page

143   "Without the Munich of 1919": In Lewis S. Feuer, *Einstein and the Generations of Science.* New York: Basic Books, 1974, p. 170.

148   "Encrypting messages": A. K. Eckert, *Physical Review Letters* 67 (1991): 661.

152   "If the packet is to be reduced, the interaction": In Edwin C. Kemble, *The Fundamental Principles of Quantum Mechanics.* New York: MacGraw-Hill, New York, 1937, p. 331.

152   "The doctrine that the world is made up": In Bernard d'Espagnat,

"The quantum theory and reality." *Scientific American*, November 1979, p. 158.

155 "That which really exists in B": Quoted by N. David Mermin, "Is the moon there when nobody looks? Reality and the quantum theory," *Physics Today*, April 1985, p. 38, from *The Born-Einstein Letters*. New York: Walker, 1971.

155 "when I consider the physical phenomena known to me": Ibid., p. 40.

156 "The extent to which an unambiguous meaning can be": Niels Bohr, "Can quantum mechanical description of physical reality be considered complete?" *Physical Review* 48 (1935): 696.

156 "If, without in any way disturbing a system": A. Einstein, B. Podolsky, and N. Rosen, "Can quantum mechanical description of physical reality be considered complete?" *Physical Review* 47 (1935): 777.

160 "There is no quantum world": Quoted in Max Jammer, *The Philosophy of Quantum Mechanics*. New York: John Wiley & Sons, 1974, p. 204, from A. Petersen, "The philosophy of Niels Bohr," *Bulletin of the Atomic Scientists*, 19 (1963): 8–14.

160 "In the experiments about atomic events": Quoted in Jammer, Ibid., p. 205, from W. Heisenberg, *Physics and Philosophy*. New York: Harper and Row, 1959, p. 160.

160 "First: I calls 'em like I sees 'em": Quoted in J. Bernstein, *Quantum Profiles*. Princeton: Princeton University Press, 1991, p. 96.

REFERENCES

ILLUSTRATION CREDITS

All illustrations not listed below are the author's own work, and were created using Mathematica and Adobe Illustrator software on an Apple Macintosh computer.

11, 12, 13: From *Astronomical Journal* 69 (1964): 75. Reproduced by permission.

18: From Henry Semat, *Fundamentals of Physics*. New York: Holt Rinehart & Winston, 1966, p. 420, fig. 24.6.

19: From L. Pearce Williams, *Michael Faraday*. New York: Da Capo Press, 1987; originally published New York: Basic Books, 1965.

29: From John Tyndall, *The Science of Sound*. New York: Citadel, 1964, p. 182, fig. 68.

37: From N. David Mermin, *Physics Today* 38, no. 4 (1985): 39, fig. 1. Reproduced by permission.

38: Courtesy of the Indiana University High-Energy Physics Group.

39: From J. W. Cronin, *Instrumentation for High-Energy Physics,* ed. F. J. M. Farley. Amsterdam: North-Holland, 1963, p. 146, fig. 5. Reproduced by permission.

40: From G. Arnison et al., *Physics Letters* (1983): 398, fig. 3. Reproduced by permission.

43: From T. Dreisch, *Handbuch der Physik,* vol. 21. Berlin: Springer Verlag, 1929, p. 184, fig. 21. Reproduced by permission.

46: From Marjorie Senechal and George Fleck, eds., *Patterns of Symmetry.* Amherst: University of Massachusetts Press, 1977, p. 59, fig. 11. Reproduced by permission.

47: From Mary H. Swindler, *Ancient Painting.* New Haven: Yale University Press, 1929, p. 45. Reproduced by permission.

48: From Organization Committee of the Exhibition of Archeological Finds of the People's Republic of China, *Chung-hua jen min kung ho kuo ch'u t'u wen wu chan lan: chan p'iu hsüan chi.* Beijing: The Wen Wu Press, 1973, pl. 60.

49: Courtesy Lee Boltin Picture Library, Croton-on-Hudson, New York. Reproduced with permission.

51: Sheet music of Johann Sebastian Bach.

52: From Owen Jones, *The Grammar of Ornament.* New York: van Nostrand Reinhold, 1982; originally published 1856. Reproduced by permission.

53: From K. Woermann, *Geschichte der Kunst.* Liepzig: Bibliographisches Institut, 1915, vol. 1, pl. 27.

54: Photo DAI Athens, NM5944. Reproduced by permission.

55: From *Historical Relics Unearthed in New China.* Beijing: Foreign Language Press, 1972, pl. 72.

56: From Jones, *Grammar of Ornament,* pl. 6. Reproduced by permission.

57: From Martin Hürlimann, *Italy.* London: Thames and Hudson, 1955, pl. 86.

58: From Ernst Haeckel, *Report on the Scientific Results of the Voyage of the H. M. S. Challenger* (1887), vol. 18, pl. 117.

59: From W. A. Bentley and W. J. Humphreys, *Snow Crystals.* New York: Dover, 1962.

60: From Dorothy Koster Washburn, *Symmetry Analysis of Upper Gila Area Ceramic Design.* Cambridge, Mass.: Harvard University Press, 1977, p. 79, fig. 107. Reproduced by permission.

61: From Branko Grünbaum and G. C. Shephard, *Tilings and Patterns.*

New York: W. H. Freeman, 1987, p. 57, fig. 2.0.1. Reproduced by permission.

62: From Senechal and Fleck, *Patterns of Symmetry*, p. 54. Reproduced by permission.

63: From Jean Favier, *The World of Chartres*. New York: Harry Abrams, 1990, p. 180. Reproduced by permission.

65: From Hermann Weyl, *Symmetry*. Princeton: Princeton University Press, 1952, fig. 16.

INDEX

Lissajous figure, 42–43
Lobachevsky, Nikolai (1792–1856, Russian), 19
Logarithms, 74
London, Fritz (1900–1954, German), 194, 199
Lorentz, Hendrik (1843–1928, Dutch), 131
Lorentz contraction, 133
Lorentz equations, 222
Lorentz transformation, 131, 232

Mach, Ernst (1838-1916, Austrian), 201
Magnetic moment, 167, 177, 187
Magnetism, 187, 189
Mass, 139, 185, 229
Maxwell, James Clark (1831–1879, Scottish), 13, 66, 67, 76, 87, 146, 220
Maxwell's demon, 76
Maxwell's equations, 116, 140, 221, 222, 232
Maxwellian distribution, 68, 69, 71, 74, 188
Mayer, Julius Robert (1814-1878, German), 58, 222
Mechanics: classical, 23ff, 144; statistical, 67, 70, 144–145, 153, 154
Mechanistic theory, 66, 67, 70
Medawar, Peter (1915–1989, Brazilian-British), 15
Meer, Simon van der (1925– , Dutch), 167
Meissner, Walther (1882–1974, German), 191
Meissner effect (Meissner-Ochsenfeld effect), 191, 199
Mendeleev, Dmitry Ivanovich (1834–1907, Russian), 146, 163
Mermin, N. David (1935– , American), 157
Mersenne, Marin (1588–1648, French), 108
Mesons, 164, 178
Michelson, Albert Abraham (1852–1931, German-American), 89
Michelson-Morley experiment, 117, 128, 129
Microwaves, 91, 101, 107, 116
Millikan, Robert Andrews (1868–1953, American), 163, 165
Mills, Robert (1927– , American), 98, 232
Minkowski, Hermann (1864–1909, Lithuanian), 132

Misener, Austin Donald (1911– , Canadian), 192
Momentum, 35, 224; conservation of, 168–169, 195, 225
Morgan, Thomas Hunt (1866–1945, American), 13
Morley, Edward William (1838–1923, American), 89
Moser, Jürgen (1928– , Swiss), 47
Muller, Hermann Joseph, (1890–1967, American), 13
Muon, 164, 170

Negentropy, 78
Neutrino, 165, 169
Neutron, 169, 177–179, 184, 194, 230
Neumann, John von (1903–1957, Hungarian-American), 52
Newton, Isaac (1642–1727, English), 11, 12, 15, 26, 79, 115, 117, 163, 220
Newton's law of cooling, 54
Newton's laws of motion, 24, 26, 127
Newtonian principle of relativity, 127
Noether, Amalie Emmy (1882–1935, German), 222
Non-Abelian gauge field theories, 98
Nonlinear equations, 119, 122, 125
Nuclear force, 185; strong, 97; weak, 97, 227
Nucleus, 163, 178

Ochsenfeld, R., 191
Onnes, Heike Kamerlingh (1853–1926, Dutch), 191, 192
Operator, 93, 148
Oppenheimer, J. Robert (1904–1967, American), 195
Order: long-range, 190, 195, 196; short-range, 190
Ozma problem, 201, 227

Parity, 226, 227
Partition function, 188
Pascal, Blaise (1623–1662, French), 27
Pasteur, Louis (1822–1895, French), 13
Pauli, Wolfang (1900–1958, Austrian), 146, 156, 163–165, 180, 184, 185, 194, 197, 227
Period, 105, 106–107
Periodic table of the elements, 164, 179, 182, 185, 220
Perpetual motion machine (Perpetuum mobile), 58, 61